Neural Networks with TensorFlow and Keras

Philip Hua

Neural Networks with TensorFlow and Keras

Training, Generative Models, and Reinforcement Learning

Apress®

Philip Hua
Guildford, UK

ISBN-13 (pbk): 979-8-8688-1019-0 ISBN-13 (electronic): 979-8-8688-1020-6
https://doi.org/10.1007/979-8-8688-1020-6

Managing Director, Apress Media LLC: Welmoed Spahr
Acquisitions Editor: Celestin Suresh John
Development Editor: Laura Berendson
Coordinating Editor: Kripa Joseph

Cover designed by eStudioCalamar

Cover image designed by Unsplash

Distributed to the book trade worldwide by Springer Science+Business Media New York, 233 Spring Street, 6th Floor, New York, NY 10013. Phone 1-800-SPRINGER, fax (201) 348-4505, e-mail orders-ny@springer-sbm.com, or visit www.springeronline.com. Apress Media, LLC is a California LLC and the sole member (owner) is Springer Science+Business Media Finance Inc (SSBM Finance Inc). SSBM Finance Inc is a **Delaware** corporation.
For information on translations, please e-mail booktranslations@springernature.com; for reprint, paper-back, or audio rights, please e-mail bookpermissions@springernature.com.
Apress titles may be purchased in bulk for academic, corporate, or promotional use. eBook versions and licenses are also available for most titles. For more information, reference our Print and eBook Bulk Sales web page at http://www.apress.com/bulk-sales.
Any source code or other supplementary material referenced by the author in this book can be found here: https://www.apress.com/gp/services/source-code.

If disposing of this product, please recycle the paper

To Kim Mai and Anson: Live a worthwhile and happy life.

Contents

About the Author

Philip Hua brings over 30 years of experience in investment, risk management, and IT. He has held senior positions as a partner at a hedge fund, led risk and IT departments at both large and boutique firms, and cofounded a successful fintech company. Alongside Dr. Paul Wilmott, he developed the CrashMetrics methodology, a crucial tool for evaluating severe market risk in portfolios. Philip holds a PhD in Applied Mathematics from Imperial College London, an MBA, and a BSc in Engineering.

About the Technical Reviewer

Shibsankar Das is currently working as a Senior Data Scientist at Microsoft. He has 10+ years of experience working in IT where he has led several data science initiatives, and in 2019, he was recognized as one of the top 40 data scientists in India. His core strength is in GenAI, deep learning, NLP, and graph neural networks. Currently, he is focusing on his research on GraphRAG, a framework that leverages RAG along with a knowledge graph to solve business problems. He has experience working in the domain of foundational research, fintech, and ecommerce.

Before Microsoft, he worked at Optum, Walmart, Envestnet, Microsoft Research, and Capgemini. He pursued a master's from the Indian Institute of Technology, Bangalore.

1 Introduction

Machine learning has become an integral part of our lives, influencing various aspects from virtual chatbots and voice assistants like Siri and Alexa to Tesla's autonomous vehicles and even the creative realms of music composition. Its applications extend to personalized advertisements, product recommendations, and investment advice. One of its remarkable achievements is DeepMind's AlphaGo, which showcases the ever-expanding capabilities of artificial intelligence (AI). In this rapidly evolving field, researchers are relentlessly pursuing the ultimate goal of AI: creating machines capable of thinking and learning, potentially rivaling human intelligence.

However, for those new to coding machine learning with Python, the plethora of online articles and concepts can be daunting. Machine learning, unlike many traditional fields, fundamentally relies on mathematics, presenting a significant challenge for those without a strong mathematical background.

For experienced software developers and data scientists, however, it is entirely possible to apply machine learning techniques to complex data challenges without delving deeply into mathematical complexities. The vast amount of available data today enables the development of highly accurate machine learning models. The key challenge lies in designing analytical workflows that extract insights from raw data, a domain where machine learning algorithms excel. Python, with its extensive numerical libraries, data preparation tools, diverse models, and robust environment, is an invaluable tool for training and scaling models for practical applications.

In this book, I aim to guide you through the fundamental concepts to advanced techniques and algorithms in machine learning. The initial chapters will clarify key concepts and definitions, helping you understand the Python code in the subsequent sections. This book is not a guide to building deep neural networks from scratch, as that requires more advanced knowledge. Instead, it equips you with the skills to effectively use machine learning algorithms and workflows in Python for complex data problems.

P. Hua, *Neural Networks with TensorFlow and Keras*,
https://doi.org/10.1007/979-8-8688-1020-6_1

Assuming a basic familiarity with Python, this book will introduce you to important libraries such as NumPy, Pandas, the deep learning framework TensorFlow, and Keras. For more sophisticated applications, we will explore commercial software APIs and models such as GPT-4 and VGG. These are built on state-of-the-art models trained on extensive datasets, offering practical solutions for specific needs rather than modifying their internal mechanics.

Wherever possible, I will utilize Google's Colab, a free service providing accessible computing resources, including GPUs and TPUs, for training machine learning models. In situations where Colab is not suitable, I will use the PyCharm IDE with a local GPU. You are encouraged to use your preferred IDE. I recommend PyCharm for its reliability and its availability in both free and paid versions. For machine learning coding, having a powerful laptop or PC with an external GPU is beneficial, as these algorithms require substantial computational resources, and a CPU alone may be insufficient for anything but the simplest tasks.

Throughout this book, I have extensively used ChatGPT-4 to explain concepts, programming syntax, and boilerplate code. While OpenAI's GPT-4 is a highly advanced tool, it is not always the perfect fit for every complex project. However, it remains an invaluable resource for assistance, offering code that, even if not always flawless, can serve as a faster starting point than writing from scratch.

Let's begin your journey into machine learning with Python.

Using Tensors

2

The advent of deep learning has revolutionized the field of artificial intelligence, allowing machines to process vast amounts of data and solve problems that were once considered too complex. As models grew larger and datasets more expansive, the computational demands of training deep learning models began to outpace the capabilities of traditional computing systems. This challenge gave rise to the need for parallel computing—a method where multiple calculations are carried out simultaneously across different processors.

Parallel computing allowed deep learning models to scale, but it also demanded new ways to efficiently represent and process data across multiple dimensions. This is where tensors come into play.

Tensors are mathematical structures that generalize the concept of vectors and matrices to higher dimensions, enabling efficient computations over large-scale data. In deep learning, tensors allow for data to be represented as multidimensional arrays, with each dimension corresponding to a particular aspect of the data—whether it be pixels in an image, time steps in a sequence, or features in a dataset.

What sets tensors apart from standard arrays used in programming languages is their optimization for parallel processing. They are designed to perform high-performance numerical computations, which makes them ideal for deep learning tasks that require intensive calculations across millions of parameters. By utilizing tensors, machine learning models can be trained on large datasets with a speed and efficiency that would be impossible with traditional data structures.

In this chapter, we will introduce tensors and their fundamental operations, laying the groundwork for building complex machine learning models. You'll learn how tensors can efficiently represent and manipulate data across multiple dimensions, making them an indispensable tool in the world of deep learning.

TensorFlow and Keras take full advantage of tensors, enabling automatic differentiation and the ability to distribute computations across multiple devices, such as GPUs and TPUs. These frameworks provide extensive support for manipulating tensors through parallel operations, allowing for faster data processing and more

P. Hua, *Neural Networks with TensorFlow and Keras*,
https://doi.org/10.1007/979-8-8688-1020-6_2

efficient memory usage. These features are key to building scalable machine learning models that handle large datasets, making tensors a powerful alternative to conventional arrays.

2.1 Tensors and NumPy

Before we start discussing tensors, some comparisons between tensors and NumPy arrays are warranted as these two are often used interchangeably in code.

Tensors, as used in TensorFlow, and NumPy arrays are both powerful tools for handling multidimensional data, and they share many similarities. However, there are key differences that make them suitable for different applications, especially in the context of machine learning and deep learning. Here are some key differences and commonality between the two libraries:

1. GPU Support: NumPy arrays are primarily designed for CPU-based computing, which means they rely on single-threaded or multithreaded CPU operations. However, deep learning models often require large-scale computations that are far more efficient when distributed across multiple processors. This is where parallel computing with GPUs becomes crucial. GPUs, with their thousands of smaller cores, are specifically built to handle many operations simultaneously, making them ideal for the matrix and tensor operations common in deep learning. Tensors, in contrast to NumPy arrays, are optimized for parallel computing. They can efficiently distribute and run computations across GPUs and TPUs. This allows for the simultaneous processing of multiple data points and operations, dramatically accelerating the training of large models. By leveraging the parallelism of GPUs, tensors make it feasible to handle the complex, large-scale calculations that deep learning requires, significantly improving efficiency and scalability.
2. Mutable: NumPy arrays are mutable, meaning that their values can be changed after they are created. In TensorFlow, tensors are typically immutable but can by mutable as well. If defined as immutable, once created, their values cannot be changed. This immutability is beneficial for optimization in computational graphs of the deep neural network.
3. Gradient Computation: Although NumPy is not inherently a deep learning library, it integrates well with various deep learning frameworks but lacks the ability to compute gradients, which are essential in training neural networks. TensorFlow's tensors are fully integrated with the TensorFlow ecosystem, allowing for automatic differentiation, which is essential for backpropagation in neural network training.
4. Eager Execution: It executes operations immediately (eager execution) in a procedural manner. TensorFlow supports both eager execution (like NumPy) and graph execution. In graph execution, operations are defined as a part of a computational graph that can be optimized and run efficiently, which is a key feature for deep learning models.

5. Broadcasting: Both NumPy array and tensor support broadcasting. Broadcasting is a mechanism that allows us to work with arrays of different shapes when performing arithmetic operations. Fundamentally, broadcasting automates the expansion of the smaller array so that it matches the shape of the larger array.

For broadcasting to work, the size of the trailing dimensions of the arrays must either be the same or one of them must be one. If this condition is not met, a broadcasting error is raised. If the two arrays are suitable, broadcasting will stretch the dimensions of the two arrays to match, as shown in the following examples:

2.1.1 Array and Scalar

When performing operations between an array and a scalar value, broadcasting allows the scalar to be "stretched" to match the shape of the array. Example: numpy_array + 3. Here, the scalar 3 is broadcasted to the shape of numpy_array.

2.1.2 Arrays with Different Dimensions

If one array has fewer dimensions than the other, it is padded with ones on its leading (left) side. Example: If A has shape (3, 5) and B has shape (5,), then B is treated as if it had the shape (1, 5). During the operation, B is broadcasted to the shape (3, 5) by repeating the same row.

2.1.3 Arrays with the Same Number of Dimensions

Broadcasting compares their shapes element-wise. A dimension of length 1 can be stretched to match the other shape. Example: If A has shape (2, 3, 1) and B has shape (1, 3, 4), the shapes are compatible. They can be broadcasted to a common shape of (2, 3, 4).

Even though tensors should be used for machine learning, there are occasions when NumPy arrays are needed, particularly when integrating TensorFlow-based models with data processing pipelines that utilize NumPy. Conversion between tensors and NumPy arrays is straightforward. To convert a NumPy array to a TensorFlow tensor, we can use tf.convert_to_tensor or simply pass a NumPy array to TensorFlow operations, as most of them can automatically convert NumPy arrays to tensors.

The following is a sample code to convert a NumPy array to a tensor:

```
import numpy as np
import tensorflow as tf

# Create a NumPy array
numpy_array = np.array([1, 2, 3, 4, 5])
```

```
# Convert to TensorFlow tensor
tensor_from_numpy = tf.convert_to_tensor(numpy_array)
# or simply
tensor_from_numpy = tf.constant(numpy_array)
```

To convert a TensorFlow tensor back to a NumPy array, we can use the .numpy() method of a tensor object. This method is available if TensorFlow is executing in eager mode, which is the default mode since TensorFlow 2.0. When converting between NumPy arrays and TensorFlow tensors, a deep copy is typically performed. This means the original and the converted objects do not share memory, and changing one will not affect the other. The .numpy() method for tensors is available in TensorFlow 2.x. In TensorFlow 1.x, the conversion process is less straightforward as it involves running a session to evaluate the tensor.

While converting, ensure that the data types are compatible; otherwise, an error will occur. TensorFlow has its own set of data types, but they are largely compatible with NumPy's data types.

The following is a sample code to convert a tensor to a NumPy array using a built-in method.

```
# Create a TensorFlow tensor
tensor = tf.constant([1, 2, 3, 4, 5])

# Convert to NumPy array
numpy_from_tensor = tensor.numpy()

# numpy_from_tensor is now a NumPy array
```

2.2 Basic Tensor Operations

2.2.1 Creating Tensors

```
import tensorflow as tf

# Creating a constant tensor
tensor_const = tf.constant([[1, 2], [3, 4]])

# Creating a variable tensor
tensor_var = tf.Variable([[1, 2], [3, 4]])
```

2.2.2 Mathematical Operations

Mathematical Operations TensorFlow supports element-wise mathematical operations, such as addition, subtraction, multiplication, and division.

```
# Element-wise addition
tensor_add = tf.add(tensor_const, tensor_var)
```

2.2.3 Reshaping Tensors

Reshaping and slicing are crucial for preparing data for various types of neural network layers. It is important to understand this section well.

Reshaping a tensor from one shape to another while keeping the total number of elements the same:

```
import tensorflow as tf

tensor = tf.constant([[1, 2, 3], [4, 5, 6]])
reshaped_tensor = tf.reshape(tensor, [3, 2])

# Original shape: (2, 3)
# New shape: (3, 2)
```

Flattening converts a tensor to a 1D tensor. This is often used when transitioning from convolutional layers to dense layers within a neural network:

```
flattened_tensor = tf.reshape(tensor, [-1])

# Original shape: (2, 3)
# New shape: (6,)
```

Adding a dimension is useful for adding a batch dimension or expanding dimensions for compatibility with certain operations, like broadcasting. When feeding a neural network data, the first dimension is the number of batches, so this operation is used very frequently:

```
expanded_tensor = tf.expand_dims(tensor, axis=0)

# Original shape: (2, 3)
# New shape: (1, 2, 3)
```

Removing dimensions of size 1. This is often used after operations like tf.reduce_sum with keepdims=True:

```
reduced_tensor = tf.reduce_sum(tensor, axis=1, keepdims=True)
squeezed_tensor = tf.squeeze(reduced_tensor)

# Original shape of reduced_tensor: (2, 1)
# New shape after squeeze: (2,)
```

Often used in sequence models like RNNs where the input needs to be reshaped into [batch_size, time steps, features]:

```
sequence_tensor = tf.reshape(tensor, [1, 2, 3])

# Original shape: (2, 3)
# New shape: (1, 2, 3) - 1 batch, 2 time steps, 3 features
```

In the context of convolutional neural networks (CNNs), channels refer to the different layers of information in an image. For example, a typical color image has three channels: red, green, and blue (RGB). Each channel contains data that represents the intensity of that color at every pixel in the image.

When a CNN processes an image, it analyzes these channels separately to learn patterns and features. The terms "channel last" and "channel first" describe how the image data is arranged in memory.

Channel last is a format where the image data is stored as (height, width, channels), with the channels (such as RGB) being the last element. Channel first refers to the format where the data is organized as (channels, height, width), meaning the channels come first in the sequence.

To switch between these formats, a technique called transposing is used, which rearranges the order of the dimensions. This step is often necessary when working with different machine learning frameworks that use different conventions for handling image data:

```
transposed_tensor = tf.transpose(tensor, perm=[1, 0])

# Original shape: (2, 3)
# New shape after transpose: (3, 2)
```

The parameter perm=[1, 0] defines the new order of the dimensions. In TensorFlow, dimensions are indexed starting from 0. So, in a 2D tensor (like a matrix), 0 refers to the rows, and 1 refers to the columns.

perm=[1, 0] means that the first dimension of the transposed tensor should be the second dimension (columns) of the original tensor, and the second dimension of the transposed tensor should be the first dimension (rows) of the original tensor. The tf.transpose operation rearranges the tensor according to the perm parameter. Essentially, it switches the rows and columns of the tensor.

For example, if we have the following 2D tensor:

[[1, 2, 3], [4, 5, 6]]

After applying tf.transpose(tensor, perm=[1, 0]), the tensor would be rearranged to

[[1, 4], [2, 5], [3, 6]]

For more complex neural network architectures, we might need to perform custom reshaping:

```
complex_tensor = tf.reshape(tensor, [3, -1, 1])

# Original shape: (2, 3)
# New shape: (3, 1, 1) - unspecified middle dimension
```

[3, –1, 1]: This is the new shape we want to give to the tensor. Here's what each element in this shape array represents:

3: The first dimension of the reshaped tensor will have a size of 3.

–1: This is a special value in TensorFlow's reshape operation. When we specify –1 for a dimension, TensorFlow automatically calculates the size of the second dimension based on the total size of the tensor and the size of the other dimensions. It ensures that the total number of elements in the reshaped tensor is the same as the original tensor. This is very useful when we are not sure about the size of a particular dimension, but we know the sizes of the other dimensions.

1: The third dimension of the reshaped tensor will have a size of 1.
For example:
[[1, 2], [3, 4], [5, 6]]

This is a tensor of shape (3,2) with three rows and two columns. After applying tf.reshape(tensor, [3, –1, 1]), the tensor would be reshaped to [[[1]], [[2]], [[3]], [[4]], [[5]], [[6]]]

which is (6,1,1). The reshape function kept the original number of elements $3 \times 2 = 6$ and created a 3D tensor with the last dimension size one as required.

2.3 Parallelism

Achieving maximum parallelism in TensorFlow generally involves leveraging its built-in capabilities for distributed computing and optimizing our code to utilize the available hardware resources effectively. This means using the built-in tensor operations as much as possible and avoiding manual computations. For example, use Tensor matrix multiplication instead of writing a loop to perform element-wise multiplication. Similarly, it pays to spend time thinking about how to manipulate arrays using TensorFlow rather than writing our own code. TensorFlow is optimized to run these operations efficiently, often leveraging parallelism on available hardware, such as GPUs or TPUs.

For example, batch matrix multiplication is a common operation in deep learning. TensorFlow's tf.matmul or tf.linalg.matmul function can be used to perform matrix multiplications in parallel across batches:

```
# Create two batched tensors of shape (batch_size, n, m)
# and (batch_size, m, p)

batch_tensor1 = tf.random.normal([64, 10, 20])
batch_tensor2 = tf.random.normal([64, 20, 30])

# Batch matrix multiplication
result_batch_matmul = tf.matmul(batch_tensor1, batch_tensor2)

# TensorFlow will perform these matrix multiplications
# in parallel across the batch
```

Achieving maximum parallelism in TensorFlow also involves leveraging its built-in capabilities for distributed computing and optimizing our code to utilize the available hardware resources effectively. Here are some steps and code examples to help write TensorFlow code for maximum parallelism:

1. Using Distributed Strategies
   TensorFlow offers several distributed strategies to parallelize the computation. The simplest one for single-machine, multi-GPU setups is the tf.distribute.MirroredStrategy. Here's an example of how to use it:

```
import tensorflow as tf

# Define the distribution strategy
strategy = tf.distribute.MirroredStrategy()

# Apply the strategy to a model creation
with strategy.scope():
model = tf.keras.Sequential([
tf.keras.layers.Dense(512, activation='relu'),
tf.keras.layers.Dense(10, activation='softmax')
])
model.compile(optimizer='adam',
loss='sparse_categorical_crossentropy',
metrics=['accuracy'])
...
```

2. Optimize Data Input Pipeline
   To ensure that the data input pipeline does not become a bottleneck, use the tf.data API for efficient data loading and preprocessing. Caching with AUTOTUNE is used frequently, although, in some cases, this does not work 100% correctly, and some other methods need to be substituted.

```
# Create a dataset
dataset = tf.data.Dataset.from_tensor_slices(
                    train_images, train_labels)

# Parallelize data preprocessing
AUTOTUNE = tf.data.experimental.AUTOTUNE
dataset = dataset.cache().shuffle(1000).batch(64).
            prefetch(buffer_size=AUTOTUNE)
```

3. Utilize GPUs Efficiently
   Ensure that TensorFlow is set up to use GPUs. TensorFlow will automatically distribute operations across available GPUs unless explicitly directed otherwise.

```
# Check if GPUs are available
print("Num GPUs Available: ",
len(tf.config.experimental.list_physical_devices('GPU')))

# TensorFlow will automatically use GPU if available
```

2.4 Machine Learning Environment

These are my personal preferences when setting up a machine learning (ML) environment. If you are familiar or happy with your current setup, then please feel free not to follow the setup below:

1. Hardware Requirements
   CPU: Buy the fastest CPU with maximum RAM. You should have at least 16 GB but ideally 32 GB or more. GPU Support: For more intensive tasks, a GPU can significantly accelerate the training of machine learning models. NVIDIA GPUs are widely supported. Ensure you have a CUDA-compatible GPU with a rating of 8+ and a minimum of 16 GB. However, my RTX 3080 graphics card only has 12 GB.
2. Operating System
   TensorFlow is compatible with Windows, macOS, and Linux. Although Windows is the most popular, using GPUs with it can be challenging due to the deprecation of certain libraries. Linux, especially Ubuntu is often preferred for deep learning tasks due to better support for GPUs and ease of environment setup.
3. TensorFlow Installation:
   Install TensorFlow using pip: pip install tensorflow. This command installs the latest version. If you have a compatible NVIDIA GPU, install the GPU version using pip install tensorflow-gpu. I use pip for all packages in this book, but it could also be done using Conda.
4. Integrated Development Environment (IDE)
   Use an IDE or text editor that you're comfortable with. Popular choices include Jupyter Notebook (great for experiments and visualization) or PyCharm. I find PyCharm community, which is free, is adequate for learning purposes.

Part I

Concepts and Basics of Machine Learning

3 How Machine Learns Using Neural Network

The Universal Approximation Theorem, first demonstrated by George Cybenko in 1989 and later refined by Kurt Hornik in 1991, forms the theoretical foundation of machine learning. This theorem states that a feedforward neural network with a single hidden layer can approximate any continuous function to arbitrary precision, given the right activation function and a sufficient number of neurons in the hidden layer.

The important consequence of the theorem is that a machine can theoretically learn to produce almost any output from the input set of data. Even in the cases of discontinuous functions, the neural network can still learn an approximation of the function.

Understanding how a machine learns is the key to understand how to create applications that can solve complex problems. A machine learning algorithm is different from a classical algorithm.

Instead of explicitly defining rules or logic to produce an output from input data, a machine learning algorithm learns the underlying patterns by itself. The algorithm takes a set of input and corresponding output samples, then iteratively adjusts its internal parameters to minimize the difference between its predicted output and the actual output. This process of minimizing the error or "loss" allows the algorithm to automatically discover the rules or logic needed to generate the desired results, without the programmer needing to specify them manually.

To make this idea concrete, let's take the following example. Suppose we are presented with a set of input numbers, 0, 1, 2, 3, 4, 5, 6, 7, 8, 9, 10, and the corresponding set of output numbers, 0, 1, 4, 9, 16, 25, 36, 49, 64, 81, and 100. The task is to predict the output number for an input number of 11.

Normally, one would write the following function to solve this problem in Python:

```
def predict(x): return x*x
print('output=',predict(11))
output= 121
```


P. Hua, *Neural Networks with TensorFlow and Keras*,
https://doi.org/10.1007/979-8-8688-1020-6_3

In this example, we guess from the list of input and output numbers that the correct mapping function may be $x \rightarrow x^2$. In this specific case, identifying the correct function was an educated guess. Had we use machine learning, the machine would learn that the correct mapping might be $x \rightarrow x^2$. The deduction process, however, would be more general and would take the following form:

- Create and configure a neural network as "supervised learning."
- Feed the neural network with the corresponding pairs of input and output numbers. For example, {0,0},{1,1},{2,4},. . .,{10,100}.
- Train the neural network.
- Use the trained neural network to predict the output using 11 as an input.

To further differentiate between traditional and machine learning algorithms, let's extend this example by adding another set of data to the problem above. Let's say that we have another set of data so that

Dataset 1: {0,1,2,3,4,5,6,7,8,9} → {0,1,4,9,16,25,36,49,64,81}

Dataset 2: {0,1,2,3,4,5,6,7,8} → {0,1,8,27,64,125,216,343,512}

This problem is now a lot more interesting. The function $y = x^2$ fits the first dataset perfectly, while the second dataset has the mapping of the form $y = x^3$. The graphs of $y = x^3$ and $y = x^2$ are shown in Figure 3-1. If we have to modify our predict function above to fit both datasets, we may decide on a curve between $y = x^3$ and $y = x^2$ which may be of the form $y = x^n$, where n is to be decided. Clearly, whichever parameter n is chosen, we cannot fit both curves at the same time, so a decision has to be made how to select n to give the "best" curve.

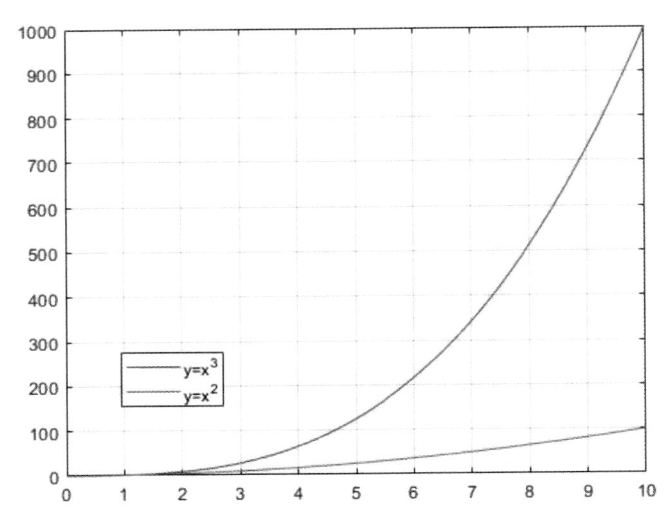

Figure 3-1 $y = x^3$ and $y = x^2$

One possible idea is to choose n to fit between the two curves to minimize the differences (or the error) between the chosen function at $x = \{1, 2, \ldots, 10\}$ to the two curves $y = x^3$ and $y = x^2$.

A similar process takes place in our machine learning algorithm. At a high level, a neural network can be considered a universal function approximator; it can approximate any given mathematical function using layers of computational "nodes" or "neurons." These nodes interact with each other to produce a function which would match the output values as much as possible from the set of input data. It does this by updating the weight of each node to reduce the value of the error function in a similar vein to what we have to do in our traditional algorithm. The main difference is that the process of choosing the weights for the nodes is iterative. The system updates the weights for the nodes iteratively for each pair of input and output data, $\{0, 0\}, \{1, 1\}, \{2, 4\}..\{10, 100\}$ and $\{0, 0\}, \{1, 1\}, \{2, 8\}, ..\{10, 1000\}$.

Well, that is the theory anyway... In practice, if we execute the code written in Keras using a simple neural network to solve the problem above, we will need to supply the model with many data points before it could approximate the function closely (Code 3-1).

The results from the neural network are shown in Figure 3-2 For the time being, there is no need to worry about the details of the code. Simply note that setting up and training a network in Python requires only a few lines of code. Also, note that the network will initially approximate a problem with straight lines; the reason for this will become apparent when we examine the internal structure later on. Of course, the power of machine learning is not limited to numerically approximating a simple function. At a high level, the neural network has just been trained to find something "similar" to the input data. As long as we can express the problem numerically, the same procedure can be used to classify images, text, sounds, translate languages, etc. We may have to use different neural network topologies to solve specific problems, but it does not invalidate the idea of using a universal numerical approximator as a "learner."

3.1 Components of a Neural Network

This section delves into the intricate components that constitute a neural network. We wish to understand each element's role and how they collectively contribute to the network's ability to learn from data and make intelligent predictions or decisions. Our exploration will cover the architecture of neurons and the building blocks of neural networks and extend to the various layers and connections that form these complex systems.

We will dissect the core components, including input layers that receive data, hidden layers that perform computations, and output layers that provide the final decision or prediction. We will also shed light on the weights and biases, crucial parameters that the network adjusts during training to improve its performance. Additionally, we will examine activation functions, which introduce nonlinear properties essential for the network's ability to solve complex problems.

```
# Initialize neural network model
model = Sequential([
Dense(10, activation='relu', input_shape=(1,)),
Dense(10, activation='relu'),
Dense(1)
])
model.compile(optimizer=Adam(), loss='mse')
initial_weights = model.get_weights()

# Define step sizes and prepare plot
step_sizes = [1, 0.1, 0.01, 0.001, 0.0001, 0.00001]
plt.rcParams['axes.grid'] = True
fig, axs = plt.subplots(3, 2, figsize=(10, 9))
fig.subplots_adjust(hspace=0.4, top=0.85)
fig.suptitle('Approximating y = x$^2$ with different
                                          iterations')


# Train and plot for each step size
for i, step_size in enumerate(step_sizes):
x = np.arange(-15, 15, step_size)
y = x**2
model.fit(x, y, verbose=0)
y_pred = model.predict(x)
model.set_weights(initial_weights)

# Plot results
axs[i // 2, i % 2].plot(x, y_pred)
axs[i // 2, i % 2].title.set_text(':,.0f
        data points'.format(len(x)))
plt.show()
```

Code 3-1 Approximating $y = x^2$ code

Furthermore, we will discuss the significance of network architecture and how different designs can be tailored to specific types of problems, from simple binary classifications to intricate tasks, like image recognition and natural language processing. We will explore how these components are trained using algorithms like backpropagation and optimized through methods such as gradient descent.

This section is designed to provide a comprehensive understanding of the components of a neural network, laying a solid foundation for grasping how these

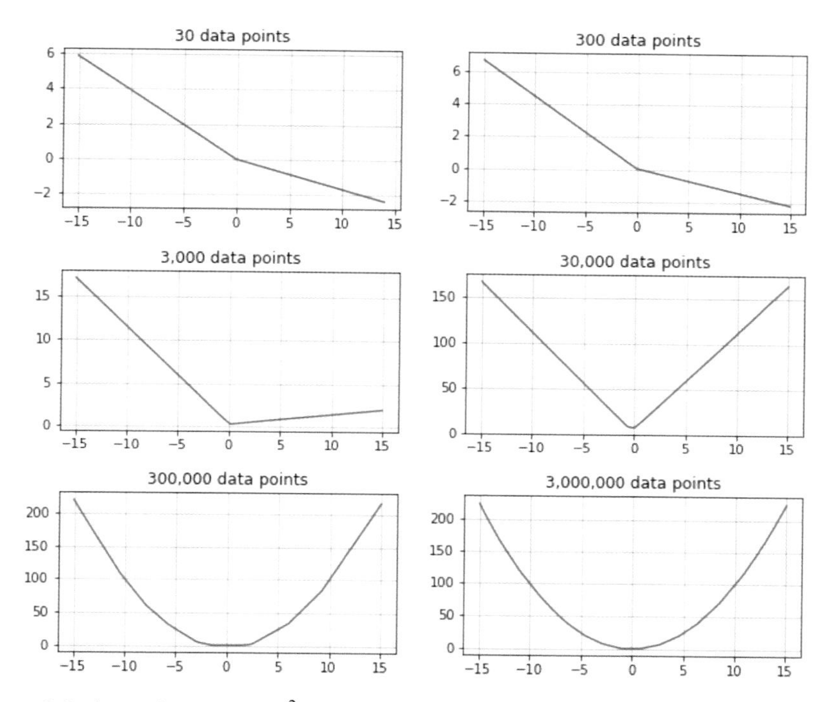

Figure 3-2 Approximating $y = x^2$ using a neural network

extraordinary systems function and are engineered to tackle some of the most challenging problems.

3.1.1 Neurons: The Building Blocks

At the core of a neural network are its neurons. These artificial neurons are inspired by the biological transmitters and receptors found in the human brain and play a pivotal role in processing and transmitting information within the network. In a typical neural network, particularly for image processing, we may encounter hundreds of thousands or millions of these neurons. Although each neuron can be thought of as the basic computing unit in a neural network, we rarely discuss individual neurons as such. Instead, we concentrate on getting the correct weights and bias terms of all the neurons in the network simultaneously:

- Input Weights: Neurons receive multiple inputs from other neurons or external data sources. These inputs are not processed equally; instead, they are associated with individual weights. These weights determine the strength of the connections between neurons. During training, the network adjusts these weights to learn and adapt to the underlying patterns in the data.

- Activation Function: After receiving weighted inputs, a neuron applies an activation function. This function introduces nonlinearity into the neural network, enabling it to capture complex relationships within the data. Activation functions transform the neuron's input into an output value, which is then passed on to subsequent neurons. Common activation functions include the sigmoid function, hyperbolic tangent (tanh), and rectified linear unit (ReLU).
- Bias Term: In addition to input weights and activation functions, each neuron typically includes a bias term. The bias term allows the network to account for possible offsets or biases in the data. It provides flexibility by allowing the network to fit the data more accurately.

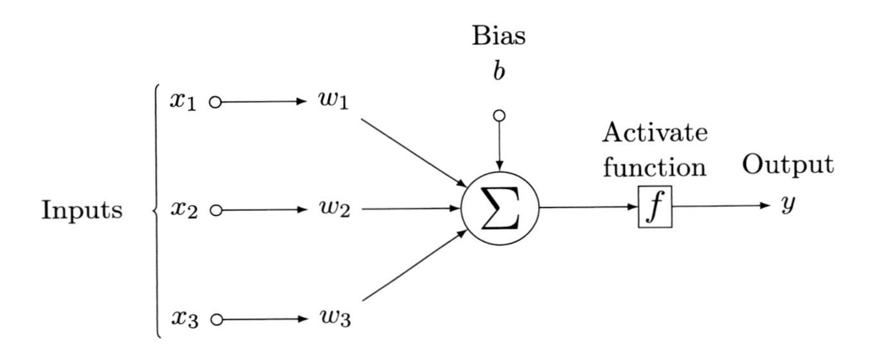

A stylized diagram for a neuron is shown in Figure 3.1.1. In this example, there are three inputs x_1, x_2, x_3. Each input can either be the data from an external source, such as the pixel values of an image, or the output from a preceding neuron. The main purpose when training the network is to find the values for the weights w_1, w_2, w_3 such that the output y is correct. In mathematical terms

$$y = f(w_1 * x_1 + w_2 * x_2 + w_3 * x_3 + b)$$

The activation function f plays an important role to introduce nonlinearity into the network; otherwise, the output is simply a linear weighted sum of the input values. This is why f is always one of the nonlinear functions shown in Figure 3-3.

The choice of activation functions, such as sigmoid, tanh, ReLU, and Leaky ReLU, depends on the neural network's architecture and the problem being addressed. These four activation functions are very popular, and most network will use one or more of these activation functions.

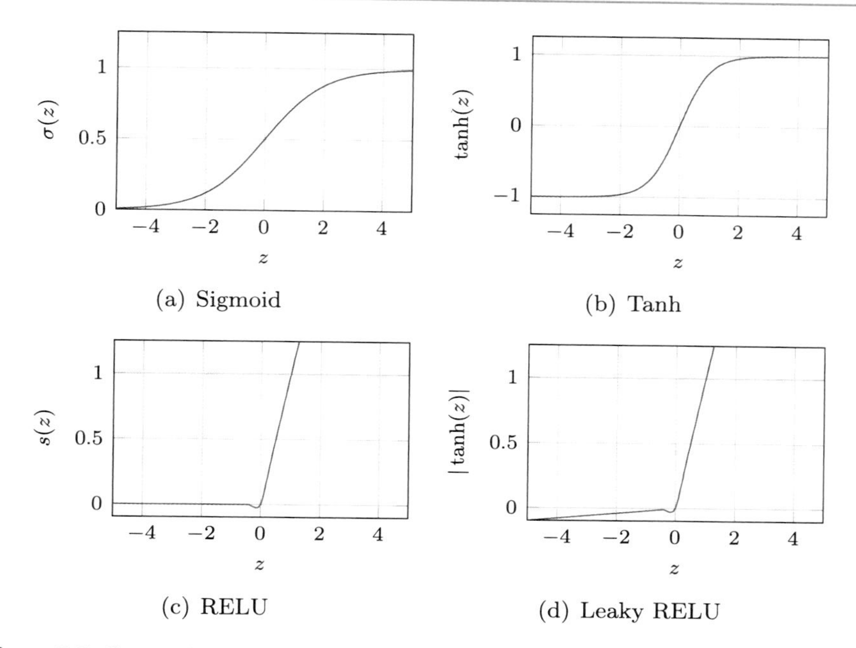

Figure 3-3 Commonly used activation functions include the sigmoid $\sigma(z)$ and the hyperbolic tangent $\tanh(z)$. More recently used activation functions are the RELU and Leaky RELU

The Sigmoid Function

The formula for the sigmoid function is $\sigma(x) = \frac{1}{1+e^{-x}}$. This function outputs values between 0 and 1, making it suitable for models that predict probabilities, such as in binary classification. It is often used in the final layer of a binary classification network to represent the probability of the predicted label being correct. For example, if we feed an image of a dog through a canine image detection network, an output value of 0.8 means that the model predicts that the image is a dog with 80% certainty.

One drawback of the sigmoid function is that it is prone to the "vanishing gradient problem."

To train this network, we need to adjust the weights of the neurons based on feedback regarding the accuracy of the predicted values compared to the actual outputs. This adjustment is done by gradients, which are essentially signals sent back through the network telling each layer how to change the weights of its neurons to improve the network's performance.

The vanishing gradient problem occurs when these gradient signals get increasingly smaller as they are passed back through each layer. Imagine if the message we are sending back through each layer gets fainter and fainter, so by the time it reaches the first layers, it is almost nonexistent. This issue arises because the earlier layers, which often capture fundamental information, receive minimal guidance on adjusting their weights.

If the layers of the network do not receive strong enough signals to learn, the entire network doesn't learn effectively. This is especially problematic in deep networks with many layers, where the early layers might not learn well, leading to poor overall performance.

The Tanh Function

An alternative to the sigmoid function, particularly popular in neural networks, is the tanh function, also known as the hyperbolic tangent. This has the formula:

$$tanh(x) = \frac{e^x - e^{-x}}{e^x + e^{-x}} \tag{3.1}$$

The tanh function is similar to the sigmoid function; however, it outputs values between –1 and 1. This means the outputs are zero-centered, which makes the training of the model easier and more efficient. It is often used in hidden layers of a neural network as it can model complex relationships more effectively than the sigmoid. However, like the sigmoid, the tanh function also suffers from the vanishing gradient problem.

The ReLU (Rectified Linear Unit) Function

The formula for the ReLU function is very simple: it is x if $x > 0$ and zero for values of $x \leq 0$. The function has gained considerable popularity, and it is the most widely used activation function for training deep neural networks and convolutional neural networks (image processing) because of its simplicity and efficiency.

ReLU (rectified linear unit) helps mitigate the vanishing gradient problem by maintaining a nonzero gradient for positive input values during backpropagation, which ensures that gradients do not shrink too rapidly as they propagate through the network.

In the ReLU function, for any input value greater than zero, the gradient is constant and equal to one. This means that during backpropagation, the weights of the network receive sufficient gradient updates, allowing the model to learn effectively. In contrast, activation functions like the sigmoid or tanh have gradients that become very small for large input values, which leads to the vanishing gradient problem as updates diminish over layers.

However, it can suffer from the "dying ReLU" problem, where neurons can sometimes become inactive and output zero for any input.

Leaky ReLU

Another popular activation function to discuss is the Leaky ReLU, similar in concept to the standard ReLU but designed to address its shortcomings. It is similar to ReLU but allows a small, nonzero gradient when the unit is not active to solve the dying ReLU problem. By allowing a small gradient α when the unit is not active, it ensures that the neurons never die. The downside is that the value of α needs to be carefully set. If it is too large, it can lead to high variance during training.

To understand concretely why the activation function is needed, we should consider the flow of signal passing through a node.

When an input signal is received at a node, it generates an output y using the following formula:

$$y = f(x_1 w_1 + x_2 w_2 + x_3 w_3 + b) \tag{3.2}$$

Here, x_1, x_2, x_3 are the outputs of all the nodes joining this node from the preceding hidden layer (or the input layer); w_1, w_2, w_3 are the weights between these interconnected nodes. The node sums the product of the weights and the input values, add a number, a (bias), and then apply the activation function f to the weighted sum to produce y.

Why does a node compute data this way?

Recall that the main purpose of a neural network is to act as a function approximator. If you have ever done a course on linear algebra, you may recall that a multiple linear regression model that approximates any set of data points with a straight line has a form of

$$y_{predict} = b + w_1 x_1 + w_2 x_2 + w_3 x_3 \ldots + w_n x_n \tag{3.3}$$

The linear regression equation has the same form as the formula for a node except that, in the case of a neuron, the activation function is applied after the weighted sum is calculated. Without the activation function, the neural network could only approximate a function with a straight line, in the same way as a linear regressor does. However, by using a nonlinear activation function, we can approximate any nonlinear target, that is, complex patterns, with less errors. The activation function also serves another purpose, that is, to activate or deactivate the next neuron. To turn off the next neuron, we just simply set the output of the activation function to zero. However, if the input to the next neuron is always zero, it would never be activated and become dead to the network.

Choosing the activation function for complex network is not an easy task and is a topic of research. In the early days, the sigmoid function was the most popular due to its nonlinearity. However, researchers found that using the sigmoid activation introduces the vanishing gradient problem for deeper networks where the weights will stop changing after some iterations, hence stopping the network from learning. The rectified linear unit activation function, ReLU, over time, became the default choice, but this too has problems as it causes dead nodes with some network configurations.

There are many activation functions available in Keras, and it is even possible to create custom activation functions using callable, but in practice, we will only need to use the following popular functions for most cases: ReLU, sigmoid, softmax, and tanh.

In Keras, there are two ways to set the activation of a layer: activations can either be used through an activation layer or through the activation argument in the dense layer itself as in the following example:

```
model = Sequential()
model.add(Dense(784, activation='relu',input_shape=(784,)))
model.add(Dense(784))
model.add(activation.ReLU(threshold=0.0))
```

The activation function is available for all forward layers, so we could also define it directly in the dense layer itself:

```
# without parameters
model.add(layers.Dense(64, activation='relu'))
or
# with parameters from keras import activations
model.add(Dense(784,lambda x:
            activations.relu(x,threshold=0.1)))
```

Note we have to use a lambda function as the first parameter of the ReLU function is a tensor or variable.

The decision of which activation function to use is tricky. The nonlinearity of the backpropagation algorithm means it is not obvious to know which function will work best in advance. For most classification problems, I would try ReLU first, then tanh or softmax for the hidden layer.

The consideration for the activation function in the output layer is more scientific as it depends on the type of result we are expecting for the problem. For multiple categorical choices, we should use *softmax*. This activation function will assign a probability to each output node, and Keras would simply choose the node with the highest probability as the most probable choice. In the MNIST example, when the network processes an image of a digit, it will assign a probability of the input image matching the ten output nodes $\{0, 1, 2, 3, 4, 5, 6, 7, 8, 9\}$.

Of course, being a probability, the sum of output node values will be one. It is also possible to use the *sigmoid* function in a multi-label classification problem, but the outputs of the nodes will not sum to one. Instead, each output represents the probability of that outcome occurring. This is useful in cases where the outcomes are not mutually exclusive. For example, a patient could have multiple symptoms of the same underlying disease.

When we have a binary classifier, that is, if the output only has two mutually exclusive outcomes, then a *sigmoid* function should be used instead. The sigmoid function, similarly to the softmax used in the case of multi-label problems, returns the probability of the input matching the output. This means that the output node will not simply be either zero or one but a real number between zero and one.

To give a concrete example of how to use a binary classifier, let us again use the MNIST database, but this time, we will only use one output node to check if the handwritten digit is eight. Of course, we would have to change the target array so that the target arrays $yTrain$ and $yTest$ elements are zero for any digit other than eight and one otherwise. The code now becomes

```
# load mnist dataset
# 60,000 images for xTrain, 10,000 for xTest
mnist = tf.keras.datasets.mnist
(xTrain, yTrain), (xTest, yTest) = mnist.load_data()
# convert the 28x28 input matrix to a vector of length 784
xTrainFlattened = xTrain.reshape(len(xTrain),784)
xTestFlattened = xTest.reshape(len(xTest),784)

#set yTrain and yTest to 1 where the digit is 8
#and 0 for others
yTrain = np.where(yTrain!=8,0,1)
yTest = np.where(yTest!=8,0,1)

# change the number of output node to 1 and
# define the loss function as binary_crossentrophy
model = Sequential()
model.add(Dense(784, activation='relu',
          input_shape =(784,)))
model.add(Dense(784,activation='relu'))
# sigmoid gives the probability of the digit is 8
model.add(Dense(1,activation='sigmoid'))
model.compile(optimizer=Adam(),
              loss='binary_crossentropy',
metrics=['accuracy'])
model.fit(xTrainFlattened, yTrain, epochs=5)
yPredict = model.predict(xTestFlattened)

# test the accuracy using the xTest
model.evaluate(xTestFlattened,yTest)
```

```
313/313 [==============================] -
2s 5ms/step - loss: 0.0278 - accuracy: 0.9904
[0.027823645621538162, 0.9904000163078308]]
```

The network identifies the figures 8 in the test set with 99% accuracy. There are 974 figures 8 in the test dataset, so there should be ten incorrectly classified images. These figures are shown in Figure 3-4. Perhaps it is more understandable why the figures 3 would be misclassified, but the figure 1 is rather more obvious, and we would expect the network to classify this correctly even when the network is not tuned.

3.1.2 Neuron Initialization

Before we start using the network, we need to initialize the weights and biases of the neurons. In Keras, the default initialization of weights varies depending on the type of layer we are using, although for common layers, such as Dense, convolutional, and RNN, the Glorot Uniform, also known as Xavier Uniform initialization, is used for the weights. This default initializer is designed to keep the scale of the gradients equal in all layers.

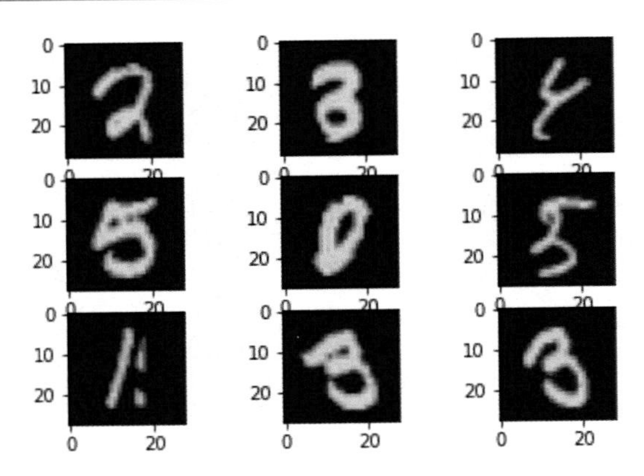

Figure 3-4 Images where NN failed to correctly identify the number

Biases are normally set to zero except for some RNN layers, such as LSTM where special initialization is required.

The choice of these defaults is based on general best practices and empirical evidence suggesting that they work well in a wide range of scenarios. However, Keras allows us to customize these initializers by specifying the kernel and bias initializer arguments in the layer constructor as shown in the example below:

```
model = Sequential([
Dense(64, input_shape=(20,), activation='relu',
kernel_initializer=RandomNormal(mean=0.0, stddev=0.05),
bias_initializer=Constant(value=0.4)),

Dense(1, activation='sigmoid',
kernel_initializer=RandomNormal(mean=0.0, stddev=0.05),
bias_initializer=Constant(value=0.4))])
```

The list of common initialization methods include

- Zero Initialization: Setting all weights to zero. This is generally not recommended because it leads to the problem of neurons learning the same features.
- Random Initialization: Setting weights to random values. This is more common, but the scale of randomness needs to be controlled. Too high values can lead to exploding gradients, while too low might lead to vanishing gradients.
- Xavier/Glorot Initialization: A popular method for initializing weights, especially in networks with sigmoid or tanh activation functions. It sets the weights based on the number of inputs and outputs of each neuron, aiming to keep the variance of outputs of each layer approximately equal.
- He Initialization: Similar to Xavier, but it's designed for layers with ReLU activation functions. It sets the weights considering only the number of inputs, which keeps the variance higher, a necessity for ReLUs.
- Orthogonal Initialization: This involves setting the weights of each layer as orthogonal matrices. It's believed to help in maintaining the stability of signals across different layers of deep networks.

The choice of initialization often depends on the type of activation function used in the network. For instance, He initialization is preferred with ReLUs, while Xavier is often used with sigmoid or tanh. Sometimes, it is beneficial to fine-tune the initialization method based on the specific characteristics of our network and the problem we are solving. Often, the best way to determine the most effective initialization method is through experimentation and observing how quickly and accurately the network learns.

3.2 Network Layers: Building a Hierarchy

While individual neurons are powerful, in practice it is the structure of the network layers that an ML engineer is more concerned with. Neural network layers serve as functional blocks, each with a specific role in processing and transforming data. The network consists of different types of layers, and their arrangement forms the network's architecture.

The topology of a neural network can get complicated, particularly for natural language processing and moving image recognition, but we will discuss them later. In the meantime, the most basic neural network is a feedforward network, which we used in our previous examples, shown in Figure 3-5.

A basic neural network has interconnected nodes, represented by circles in Figure 3-5 in three layers: an input layer, a hidden layer, and an output layer. Deep neural networks may have several hidden layers interconnected with each other before reaching the output layer. In TensorFlow and Keras, the feedforward layer is referred to as a dense layer. If we refer to the code in Code 3-1, we can see that a dense layer is defined with the number of nodes and an activation function:

```
model.add(Dense(10, activation='relu'))
```

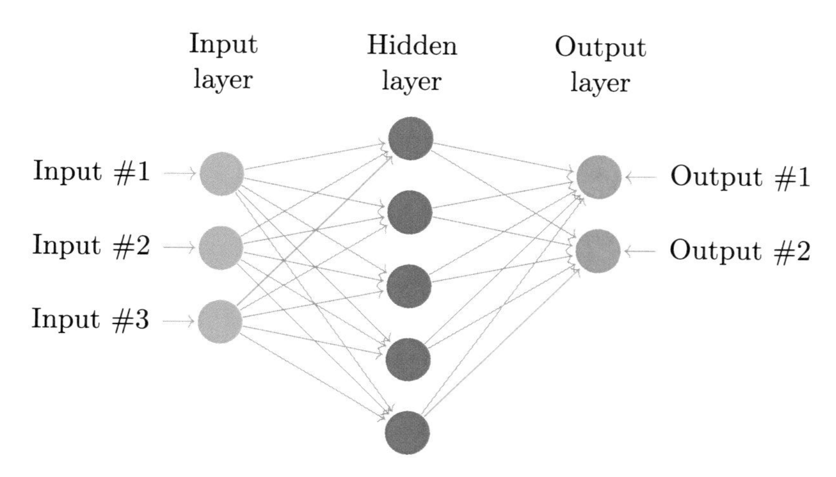

Figure 3-5 A three-layer feedforward network with three input, five hidden, and two output nodes

The hidden layers may have millions of nodes capable of mapping any input type to any output type. Of course, the more nodes we have, the more data and time are needed to train the network. A reasonably complicated deep learning network may need millions of data points rather than thousands or hundreds of thousands, so it is normally best practice to keep the network structure as lean as possible.

In a feedforward network, information flows from left to right, that is, from the input layer to the hidden layer(s) and then finally to the output layer. A number, called weight, represents a connection between two nodes, shown as an arrow in Figure 3-5.

This means that each node in the hidden or output layer receives the output weights of the nodes from the preceding layer as input, does some predefined computation, and then passes the result as a number, or weight, to the next hidden or output layer. This weight is an important number. An "important" node will have a higher weight, while a node which produces a zero weight all the time is useless and is called a dead node.

A feedforward network uses a feedback process to improve its prediction over time. Each time we supply the network with a new input, it predicts the output value using the weights and then compares how close the predicted value is to the actual output value using a *loss function*. The network then adjusts the weights of the nodes in the hidden layer(s) using an algorithm called *backpropagation* (or *backprop* for short) to reduce the error. This process is repeated until all the inputs are processed and the weights are updated accordingly.

It should be clear that three components are very important for a feedforward network: the activation function, the loss functions, and the feedback algorithm. Together they determine how a network gets to learn quickly and correctly.

3.3 The Optimizer and the Loss Function

Other than the activation function, we need to discuss the optimizer and the loss function for a neural network. We may see this in the previous code as

```
model.compile(optimizer=Adam(), loss='binary_crossentropy',
metrics=['accuracy'])
```

Both of these components are essential to using a neural network. Unfortunately, we will need to have a mathematical background to understand the concepts properly. However, I will illustrate the ideas with analogies as much as possible without resorting to complicated mathematical formulas.

The key idea here is that we need to "train" our network to achieve what we want it to do. For each predicted output, we need to (1) identify how close we are to the actual output and (2) change the weights of the neurons after each predicted output to (hopefully) inch the network toward the correct solution. Step 1 is done using the loss function, while it is the job of the optimizer via the backpropagation algorithm in step 2 to improve the accuracy of the network.

3.3.1 The Loss Function

The loss function provides the backpropagation algorithm of an estimate of the size of the error in the current state of the model. It provides the backpropagation optimization algorithm with a single number to minimize. The smaller the error, the better the model is at fitting the training data. However, for reasons which will be discussed later, it is not necessarily the case that we would want the lowest error during the training process as the model may not perform as well during testing or production because it has overfitted the training data.

In Keras, there are three types of loss functions. Which one we should use depends on the type of problem under consideration. For regression, that is, approximating real valued functions, we should use a regression loss. With problems where we need to determine the probability of occurrence, then use one of the probabilistic loss functions. The third type, called hinged losses, is specifically designed for "maximum-margin" classification problems. Underlying any loss function is the concept of proximity measure, which gives a mathematical metric of how close, or equivalently how similar, two objects are to each other.

All losses are available in Keras via a class handle and a function handle. The class handles enable us to pass configuration arguments to the constructor as follows.

Using a Function Handle

When we use a function handle, we typically specify the loss function by its string name when compiling the model:

```
from tensorflow.keras.models import Sequential
from tensorflow.keras.layers import Dense

# Define a simple model
model = Sequential([
Dense(64, activation='relu', input_shape=(10,)),
Dense(1)
])

# Compile the model using a function handle for the loss
model.compile(optimizer='adam', loss='mean_squared_error')
```

In this case, the mean squared error is the string identifier for the mean squared error loss function.

Using a Class Handle

When using a class handle, we instantiate a loss class with any required configuration parameters and then pass this instance to the compile method of the model. This approach provides more flexibility, as we can pass additional parameters to the loss function if the loss function supports such parameters. Here is an example using the MeanSquaredError class:

```
from tensorflow.keras.models import Sequential
from tensorflow.keras.layers import Dense
from tensorflow.keras.losses import MeanSquaredError
```

```
# Define a simple model
model = Sequential([
Dense(64, activation='relu', input_shape=(10,)),
Dense(1)])

# Instantiate the class with desired configuration
mse_loss = MeanSquaredError(reduction='auto',
                name='mean_squared_error')

# Compile the model using the class handle for the loss
model.compile(optimizer='adam', loss=mse_loss)
```

3.3.2 Loss Function for Regression

Regression problems are a set of problems where the model predicts a real value as output. For example, predicting house prices, company profits, or future stock prices. Keras provides both class functions to instantiate objects and set parameters, as well as loss functions that can be called directly. In either case, the mean squared error is by far the most popular loss function used to solve regression problems, while cosine similarity is frequently used in natural language processing to measure how similar words are. We will discuss these two functions below. If you are interested in learning more, here is the complete list of loss functions implemented in Keras: https://keras.io/api/losses/.

3.3.3 Mean Squared Error

The mean squared error (MSE) is frequently used in statistics and machine learning to measure the average of the squares of the errors, essentially quantifying the difference between predicted and actual values. Let's suppose that we have the actual values for our problem as

$$Y = \{10.2, 11.0, 12.5, 1.0, -6.5, 0.0\}$$

and the predicted values from our network are

$$Y_{predict} = \{9.9, 12.2, 12.5, 0.0, 1.0, 1.0\}$$

then the MSE is given by

$$\begin{aligned} MSE = \frac{1}{6}[(10.2 - 9.9)^2 + (11.0 - 12.2)^2 + (12.5 - 12.5)^2 + \\ (1.0 - 0.0)^2 + (-6.5 - 1.0)^2 + (0.0 - 1.0)^2] = 9.963333 \end{aligned} \tag{3.4}$$

As each term in the MSE formula is squared, the error function is always positive, ensuring that the function focuses on the magnitude of errors. It becomes zero only when the predicted values match the target values exactly. Calculating MSE in Keras is straightforward and gives the same result as above.

```
import numpy as np
import tensorflow as tf

y= [10.2,11,12.5,1,-6.5,0]
yPredicted = [9.9,12.2,12.5,0,1,1]
mse = tf.keras.losses.MeanSquaredError()
print(mse(y,yPredicted).numpy())
```

```
9.963333
```

This loss function can be specified in the machine learning function *compile* as follows:

```
model.compile( loss=tf.keras.losses.MeanSquaredError())
```

3.3.4 Cosine Similarity

In contrast to regression, which deals with numerical predictions, cosine similarity is used to measure the similarity between two vectors, commonly in applications involving text or sequences. For example, suppose we want to compare two similar sentences in the popular nursery rhymes "Humpty Dumpty":

"Humpty Dumpty sat on a wall" and "Humpty Dumpty had a great fall"

One way to do this is to vectorize each sentence using 0 and 1 to indicate if a word is in the dictionary as shown in Figure 3-6. Once vectorized, the cosine of the angle α between the two vectors is a number representing how close (or similar) they are to each other. The cos() function has values ranging from –1 to 1. An angle of zero degrees (cos(0) = 1) indicates two vectors pointing in the same direction, signifying maximum similarity. Conversely, cos(180 degrees) equals –1, indicating vectors in opposite directions, signifying maximum dissimilarity, as shown in Figure 3-7c.

The advantage of using the cosine similarity metric is that it is very easy and fast to calculate. We simply compute the dot product of the two vectors and divide by their lengths as follows:

$$cos\alpha = \frac{A.B}{||A||||B||}$$

where $||A||$ denotes the length of vector A. For example, the angle between the vector $\{3, 4, 5\}$ and $\{-3, 5, 4\}$ is

$$cos\alpha = \frac{3.(-3) + 4.5 + 5.4}{\sqrt{3^2 + 4^2 + 5^2}.\sqrt{(-3)^2 + 5^2 + 4^2}} = 0.62 \tag{3.5}$$

Figure 3-6 Vectorizing sentences

Word	Vector a	Vector b
Humpty	1	1
Dumpty	1	1
sat	1	0
had	0	1
a	1	1
on	1	0
great	0	1
fall	0	1

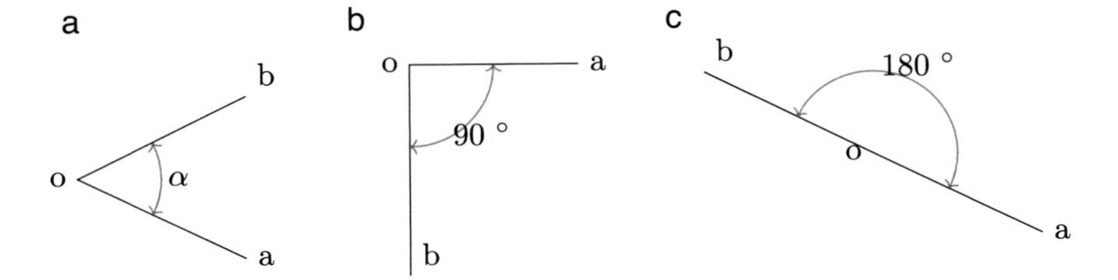

Figure 3-7 Cosine similarity between two vectors

Keras has a built-in function to perform the same calculation:

```
import numpy as np
import tensorflow as tf

y= [3.0,4.0,5.0]
yPredicted = [-3.0,5.0,4.0]
cosineLoss = tf.keras.losses.CosineSimilarity()
print(cosineLoss(y,yPredicted).numpy())
```

```
-0.6199999
```

We can use this in the *compile* function as follows:

```
model.compile(optimizer='Adam',
loss=tf.keras.losses.CosineSimilarity(axis=1))
```

We may notice the parameter axis=1. This tells the algorithm to calculate the cosine similarity of the vector to the feature axis. We will explore the idea of a feature axis and what it means later on when we discuss the use of a convoluted neural network (CNN) for image processing.

3.4 Probabilistic Losses

For categorical problems, that is, for problems which require us to identify classes of objects, for example, images of dogs, cats, birds, fish, etc., we may require the neural network to output the probabilities of the input object belonging to each class. Keras provides many loss functions to do this, but the most popular options are binary, categorical cross-entropy, and sparse categorical cross-entropy classes, which we will discuss below.

3.4.1 Binary Cross-Entropy

Use this entropy function for classification problems where the output has only two outcomes, for example, a picture of a dog or cat, the word "happy" or "sad." For a binary classification model, the output is the probability of a sample belonging to one of the two classes, usually represented as 0 or 1. The binary cross-entropy loss function measures how far the model's predictions are from the actual labels. For a single sample, it is defined as

$$L(y) = -y\log(p) - (1-y)\log(1-p) \tag{3.6}$$

For many data points, the loss function is simply the arithmetic average of the total loss for all data points. p is the predicted probability coming from our neural network. Note that since y can only be 1 or 0, only one part of the equation can be activated. When $y = 1$, $L(1) = -\log(p)$. If the predicted probability $p = 1$, the $L(1) = 0$, so there is no loss. If the probability p is ≈ 0, $-\log(p) >> 1$, so the loss value is very high. The same logic works in the case of $y = 0$ for $L(0)$.

If we use Keras to calculate binary cross-entropy for our neural network, it follows the same format as before. That is, we need to set the parameter *loss* in the *compile* function.

```
lossFunction = BinaryCrossentropy()
model.compile(loss=lossFunction,optimizer='adam',
activation='sigmoid',metrics=['accuracy'])
```

Use the activation function *sigmoid* for a binary cross-entropy loss function.

3.4.2 Categorical Cross-Entropy

This is the same as the binary cross-entropy loss except that it is used for multi-class outcomes instead of binary. For example, the output could be a picture of a cat, a dog, a bird, or others. If we use Keras categorical cross-entropy, we will need to vectorize the output first so that each output is either 1 or 0. This is done by using one-hot encoding. Simply put, one-hot encoding assigns a unique one and zero vector for each category as shown in Figure 3-8.

Figure 3-8 One-hot encoding for the three categories: red, yellow, and blue

Red	Yellow	Blue
1	0	0
0	1	0
0	0	1

The *softmax* is a suitable activation function for categorical cross-entropy. The softmax function converts the output of the neural network into probability distributions, ensuring that the sum of the probabilities of all output nodes is equal to one.

3.4.3 Sparse Categorical Cross-Entropy

The sparse categorical cross-entropy function is similar to categorical cross-entropy, with the key difference being that it does not require one-hot encoding of output labels. Instead, the labels can be directly assigned as integers. For example: $red = 1$, $yellow = 2$, $blue = 3$. The advantage of using sparse cross-entropy is that it saves time and memory as each class is an integer and not a whole vector.

3.5 Network Optimizer

In machine learning, optimization refers to the process of iteratively refining the model parameters to improve its accuracy and reduce error.

As mentioned, this is done using an optimizer in the backpropagation process. The main objective of the optimizer is to minimize the value of the loss function after each iteration by changing the weights of the neurons. It achieves this iteratively using a numerical search algorithm that estimates the next set of weight values based on the loss function. The new estimated weights should reduce the loss over iterations, but there is no guarantee that it will converge or, if it converges, that it will converge to a global maximum.

The algorithm to optimize the weights is normally based on one of the popular algorithms called gradient descent. To understand the intuition behind this algorithm, imagine that we are lost on a hill walk and it is now dark. One strategy to find the bottom of the hill is to explore the immediate surrounding and walk a few steps in the direction of the steepest slope, then check for direction again and repeat until we get stuck in some hole or got the bottom of the hill. The strategy requires two parameters: the number of steps before we check for direction (or equivalently, the size of each step if we check after every step) and the direction of descent at each step.

Choosing a small step size would slow convergence. Conversely, a step size that is too large could mean that we miss the nuances of the ground topology and end up walking randomly up and down the hill. The optimal step size depends on each application, but one thing which would help the optimizer, in machine learning and not in real life, is to normalize the data so that a similar step size could be used for many problems. We will discuss this point in more detail in later sections.

The direction of the steepest gradient for the optimizer aligns with the local gradient of the loss function, which is why we need to choose a loss function for every problem. Different variants of the gradient descent algorithm may improve the process by adapting the step size depending on the steepness of the gradient. They may also introduce a random walk now and again to avoid getting stuck in a hole, but the basic idea remains the same.

There are several optimizers to choose in Keras, but they fall into two categories: basic and adaptive gradient descent algorithms. We can read the different types on the Internet if we are curious, but in practice, the two most used ones are mini batch gradient descent and Adam.

We often use Adam first for our training, as it is considered one of the best among the adaptive optimizers. However, the mini batch gradient is also worth trying if convergence is slow or if there are memory limitations. The mini batch term refers to taking the average of the loss function after a batch of input data before updating the weights. There is no magic formula for the batch size, but we will discuss some consideration for choosing the value for this parameter later on.

An important hyperparameter for an optimizer is the learning rate and its schedule. The learning rate is a critical hyperparameter that controls how much the weights in the network are adjusted with respect to the loss gradient. A smaller learning rate makes the network learn slower, but it can help the network reach a better or more precise final performance. A larger learning rate makes the network learn faster, but it can overshoot the optimal values. Keras also provides a facility to change the active learning rate (or its schedule) via the use of a callback function.

To define the optimizer, use

```
adam = Adam(learning_rate=0.001)
```

The model is then set up to use the optimizer by compiling

```
model.compile(optimizer=adam, loss='binary_crossentropy',
        metrics=['accuracy'])
```

In Keras, learning rate schedules are mechanisms used to adjust the learning rate during training, allowing for dynamic adaptation based on the model's performance. This can lead to better performance of the model, as it allows for a more refined tuning of the optimization process. To illustrate, common types of learning rate schedules include time-based decay, step decay, and exponential decay, each with its own strategy for adjusting the learning rate. Here are examples of each.

Time-Based Decay

```
from tensorflow.keras.models import Sequential
from tensorflow.keras.layers import Dense
from tensorflow.keras.optimizers import Adam
from tensorflow.keras.callbacks import LearningRateScheduler
import numpy as np
```

```
# Sample model
model = Sequential([Dense(64, activation='relu',
                          input_shape=(20,)),
                    Dense(1, activation='sigmoid')])

# Initial learning rate
initial_learning_rate = 0.01

# Define the scheduler function
def scheduler(epoch, lr):
     decay = 0.1
     return initial_learning_rate / (1 + decay * epoch)

# Compile the model with the initial learning rate
model.compile(
        optimizer=Adam(learning_rate=initial_learnin_rate),
    loss='binary_crossentropy', metrics=['accuracy'])

# Fit the model with the learning rate scheduler
model.fit(X_train, y_train, epochs=100,
            callbacks=[LearningRateScheduler(scheduler)])
```

Step Decay

Step decay reduces the learning rate by a factor every few epochs.

```
# Define the scheduler function for step decay
def step_decay(epoch):
     initial_lr = 0.01
     drop_rate = 0.5
     epochs_drop = 10.0
     return initial_lr * np.power(drop_rate,
               np.floor((1+epoch)/epochs_drop))

# Compile and fit as before
model.compile(optimizer=Adam(learning_rate=0.01),
  loss='binary_crossentropy',
      metrics=['accuracy'])
model.fit(X_train, y_train, epochs=100,
     callbacks=[LearningRateScheduler(step_decay)])
```

Exponential Decay

```
# Define the scheduler function for exponential decay
def exponential_decay(epoch):
     initial_lr = 0.01
     k = 0.1
     return initial_lr * np.exp(-k*epoch)

     # Compile and fit as before
     model.compile(optimizer=Adam(learning_rate=0.01),
       loss='binary_crossentropy',
          metrics=['accuracy'])
     model.fit(X_train, y_train, epochs=100,
     callbacks=[LearningRateScheduler(exponential_decay)])
```

3.6 Generalization Errors

If a network performs well on the training set but generalizes badly, it is *overfitting* the data. A network might overfit if the training set contains accidental regularities in the input data. For instance, in our MNIST training dataset, if the handwritten digit was from a single person, then any quirks in the way a digit is written could be taken as gospel truth in our trained network, fooling it to failing to distinguish the handwriting from other people. Equally, suppose we have a diverse handwritten dataset from millions of people, a simple network with few neurons will not have the capacity to remember the information in its training data. In other words, the number of neurons and possibly the architecture of the network are insufficient to capture the complexity of the data. When a network fails to learn this way, it is *underfitting* the data. An overfitting error is also known as *variance* error. A network with a high overfitting error is said to exhibit a high variance. Similarly, an underfitting error is known as *bias* error. A high level of underfitting means a high level of bias. We can deduce from the argument above that over- and underfitting errors are related. We reduce the overfitting error by reducing the capacity of the network or adding additional training examples. However, if a model is underfitting, adding data does not help. No matter how much we reduce these two types of error, we should understand that for real-world problems, the network can never predict with 100% accuracy, so it is a compromise of accepting a low error rate rather than achieving perfection.

A learning curve, illustrated in Figure 3-9, is a graphical representation of the relationship between a model's performance and a measure of experience, such as the number of training instances or iterations. It visualizes how effectively the model learns over time and highlights potential issues like overfitting or underfitting.

Keras and other machine learning languages have tools to measure and reduce generalization errors, which we will discuss next. It is important to keep in mind that the overall objective of these tools and procedure is to make the best use of the training dataset and reduce the generalization errors.

3.7 TensorBoard

It is imperative that during training, we keep track of how well a model is doing. For very simple model, it may suffice to include debugging statements to show the average loss per epoch. A better and more informative way to track the performance of the model is to use TensorBoard.

TensorBoard, provided by TensorFlow, Google's open source machine learning framework, is a visualization toolkit designed for machine learning experimentation. It facilitates the inspection and understanding of machine learning workflows. It offers a suite of web applications that help us visualize and understand our TensorFlow runs and graphs. TensorBoard is particularly useful in the training and fine-tuning of neural networks.

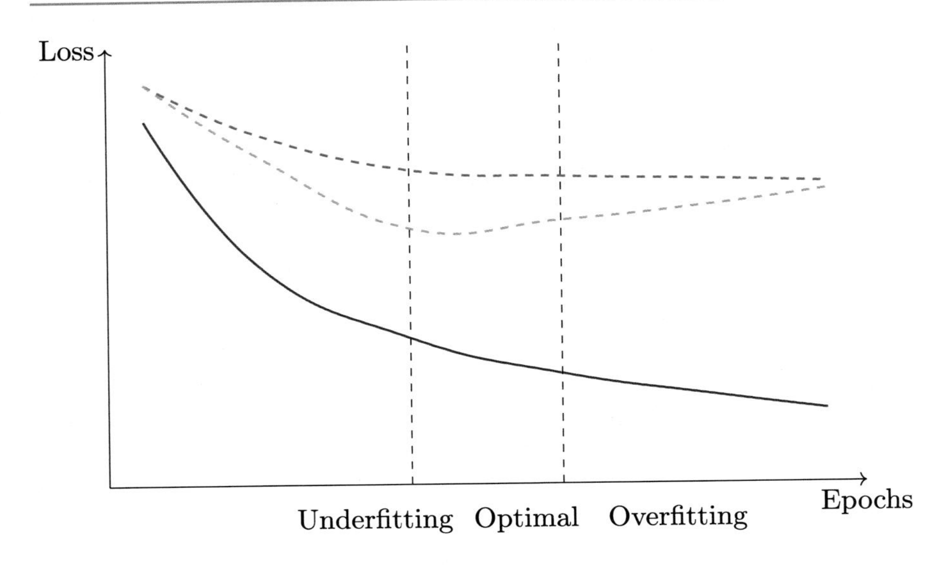

Figure 3-9 The learning curve

Setting up and using TensorBoard is straightforward, typically involving the integration of a few lines of code within the TensorFlow training script to enable logging and visualization. It does not impose a significant amount of performance penalties, so it is a practical tool to use in many instances. I personally prefer to first develop and test the model and then integrate TensorBoard for detailed monitoring during actual training runs.

Key Features of TensorBoard

- Visualization of Metrics: It allows us to track and visualize metrics such as loss and accuracy during the training process. We can see these metrics in real time, which helps in understanding how our model is performing and when it begins to overfit or underfit.
- Graph Visualization: TensorBoard provides a way to visualize our model architecture. This can help in understanding the TensorFlow graph, observing how tensors flow through the graph, and debugging if necessary.
- Viewing Histograms of Weights and Biases: We can see the distribution of weights and biases across different layers in the network over time. This can give insights into how the network is learning.
- Projector for Embeddings: TensorBoard includes a tool for visualizing high-dimensional embeddings. This feature is particularly useful for tasks like word embeddings in natural language processing.

- Image and Audio Visualization: If we are working with image or audio data, TensorBoard can show actual images or play audio directly within the dashboard, which can be useful for monitoring the outputs of our model.
- Hyperparameter Tuning: With the HParams dashboard, we can visualize hyperparameter tuning experiments with Keras Tuners (or another tuning library). We can record the hyperparameters (like learning rate, number of layers) and metrics (like loss, accuracy) and then compare different runs to see which hyperparameters work best.
- Performance Profiling: TensorBoard also offers tools to profile the model, helping us understand where the bottlenecks in computation are and how efficiently our model is running.

To use TensorBoard, we typically start by modifying our TensorFlow code to write log files containing the metrics, embeddings, etc., to a specified directory. Then we launch TensorBoard and point it to this log directory and visualize in the web browser.

Here are some basic examples of how to use TensorBoard with a TensorFlow/Keras model:

Metric Visualization

```
import tensorflow as tf
from tensorflow.keras.layers import Dense, Flatten, Conv2D
from tensorflow.keras.callbacks import TensorBoard
import datetime

# Prepare dataset (example with MNIST)
mnist = tf.keras.datasets.mnist
(x_train, y_train), (x_test, y_test) = mnist.load_data()
x_train, x_test = x_train / 255.0, x_test / 255.0

# Add a channels dimension
x_train = x_train[..., tf.newaxis]
x_test = x_test[..., tf.newaxis]

# Build the model
model = tf.keras.models.Sequential([
Conv2D(32, 3, activation='relu', input_shape=(28, 28, 1)),
Flatten(),
Dense(128, activation='relu'),
Dense(10)
])

# Compile the model
model.compile(optimizer='adam',
loss=tf.keras.losses.
    SparseCategoricalCrossentropy(from_logits=True),
metrics=['accuracy'])

# Set up the TensorBoard callback
log_dir = "logs/fit/" +
```

```
    datetime.datetime.now().strftime("%Y%m%d-%H%M%S")
tensorboard_callback = TensorBoard(log_dir=log_dir,
     histogram_freq=1)

# Train the model
model.fit(x=x_train,
y=y_train,
epochs=5,
validation_data=(x_test, y_test),
callbacks=[tensorboard_callback])
```

Graph Visualization
The model architecture is automatically logged by TensorBoard. We can view it in the Graphs tab.

Histograms of Weights and Biases
The histogram freq=1 parameter in the TensorBoard callback logs the distribution of weights and biases. These can be viewed in the Histograms tab in TensorBoard.

Embedding Visualization
For embedding visualization, we need to have an embedding layer in our model and use the TensorBoard callback with an embedding layer specified.

```
# Assuming 'model' has an embedding layer
tensorboard_callback = TensorBoard(log_dir=log_dir,
histogram_freq=1, embeddings_freq=1)
```

To visualize images, we need to modify the TensorBoard callback:

```
file_writer = tf.summary.create_file_writer(log_dir + '/img')
with file_writer.as_default():
tf.summary.image("Training data", x_train, step=0)
```

Performance Profiling
```
tensorboard_callback = TensorBoard(log_dir=log_dir,
histogram_freq=1, profile_batch='500,520')
```
The HParams dashboard, a feature within TensorBoard, offers a specialized interface to visualize and analyze the results of hyperparameter tuning experiments, aiding in the selection of the most effective model configurations. It allows us to interactively compare the performance of different sets of hyperparameters, making it easier to identify the most effective configurations for our machine learning models.

To use the HParams dashboard, we need to log hyperparameters and metrics during our model's training process. Essentially, we need to loop through the set of parameters and record the results for each run using TensorBoard as shown below:

```
from tensorboard.plugins.hparams import api as hp
```

```
HP_NUM_UNITS = hp.HParam('num_units', hp.Discrete([16, 32]))
HP_DROPOUT = hp.HParam('dropout', hp.RealInterval(0.1, 0.2))
HP_OPTIMIZER = hp.HParam('optimizer',
                hp.Discrete(['adam', 'sgd']))

METRIC_ACCURACY = 'accuracy'

with tf.summary.create_file_writer(
                    'logs/hparam_tuning').as_default():
hp.hparams_config(
     hparams=[HP_NUM_UNITS, HP_DROPOUT, HP_OPTIMIZER],
     metrics=[hp.Metric(METRIC_ACCURACY,
        display_name='Accuracy')],)

def train_test_model(hparams, session_num):
    model = tf.keras.models.Sequential([
     tf.keras.layers.Dense(hparams[HP_NUM_UNITS],
         activation='relu'),
     tf.keras.layers.Dropout(hparams[HP_DROPOUT]),
     tf.keras.layers.Dense(10, activation='softmax')])
model.compile(
optimizer=hparams[HP_OPTIMIZER],
loss='sparse_categorical_crossentropy',
metrics=['accuracy'],)

# Run with the hparams
model.fit(x_train, y_train, epochs=10) # example values
_, accuracy = model.evaluate(x_test, y_test)
return accuracy

session_num = 0

for num_units in HP_NUM_UNITS.domain.values:
for dropout_rate in (HP_DROPOUT.domain.min_value,
        HP_DROPOUT.domain.max_value):
for optimizer in HP_OPTIMIZER.domain.values:
hparams =
HP_NUM_UNITS: num_units,
HP_DROPOUT: dropout_rate,
HP_OPTIMIZER: optimizer,

run_name = "run-%d" % session_num
print('--- Starting trial: %s' % run_name)
print(h.name: hparams[h] for h in hparams)
accuracy = train_test_model(hparams, session_num)
session_num += 1

# Log an hparams summary with the metrics.
with tf.summary.create_file_writer('logs/hparam_tuning/'
                + run_name).as_default():
hp.hparams(hparams) #record the values used
tf.summary.scalar(METRIC_ACCURACY, accuracy, step=1)
```

Run TensorBoard and navigate to the HParams tab, then use

```
tensorboard --logdir logs/hparam_tuning
```

. Open the browser and go to http://localhost:6006/. In the HParams dashboard, we can

- View tables and visualizations of runs
- Filter and sort based on hyperparameters and metrics
- Explore parallel coordinates and scatter plot views to analyze relationships between hyperparameters and model performance

3.8 Using TensorBoard in Colab

When using TensorBoard within Google Colab, the procedure remains largely the same as outlined earlier. However, it's important to enable "allow third-party cookies" in the browser settings to avoid encountering a "403 Error."

```
%tensorboard --logdir logs_directory
```

Use Colab's file manager to locate the path of the logs and use that path for the `logs_directory` variable. The TensorBoard log contains enough of data for other metrics such as confusion matrix to be calculated, but we will have to search add-ins on the Internet.

Note that if we wish to view the model graph, then select the Graphs tab. By default, the model is shown inverted with data flowing from bottom to top. The default graph is slightly different to the model produced by Keras's function `plot_model()`, but it is perhaps more intuitive than the standard view, but if we need to see the same model as Keras, then select the conceptual graph.

Sometimes, we may run into an error with TensorBoard blocking the port 6006 because it has been occupied. If this is the case, we will need to kill the existing process on the port. Unless we have a Colab pro subscription which gives access to the terminal app, one way to find the correct PID to kill is to create an instance of a terminal using the following commands:

```
!pip install colab-xterm
%load_ext colabxterm
%xterm
```

Type `lsof -i:6006` in the terminal to bring up the PID number and use the displayed PID to kill the process with the command `kill PID`.

Network Layers

4

In the context of machine learning, and more specifically in neural networks, “layers” refer to various levels or stages of processing units. These layers are fundamental in extracting and transforming features from input data, each serving a distinct functional purpose in the learning process. Each type of layer is designed to perform a specific kind of operation on its input data. Here are some common types of layers we will encounter in machine learning models.

These are the most basic type of layer in neural networks, where each neuron is connected to every neuron in the preceding and subsequent layers. They are typically used for learning nonspatial hierarchies of features.

4.1 Dense (Fully Connected) Layers

A dense layer, characterized by its fully connected nature, can be versatilely used as an input layer, a hidden layer, or an output layer in a neural network, depending on the specific architecture and requirements of the model. We define the dense layer for the three cases below:

```
from keras.models import Sequential
from keras.layers import Dense

# Create a Sequential model
model = Sequential()

# Adding the input layer
# Assume input_dim is the size of the input features
model.add(Dense(units=64, activation='relu',
        input_dim=100))
```


P. Hua, *Neural Networks with TensorFlow and Keras*,
https://doi.org/10.1007/979-8-8688-1020-6_4

```
# Adding a hidden layer
model.add(Dense(units=32, activation='relu'))

# Adding the output layer
# Assuming it's for a binary classification problem
model.add(Dense(units=1, activation='sigmoid'))
```

In the example above

- Number of Units (Neurons): The units parameter in a dense layer specifies the number of neurons. The appropriate number of units can vary depending on the complexity of the task and is generally determined through experimentation.
- Activation Function: ReLU is commonly used in hidden layers because it helps with the vanishing gradient problem and allows the model to learn complex patterns. The sigmoid function in the output layer is typical for binary classification.
- Input Dimensions: The "input_dim" parameter is crucial for the first layer in a sequential model as it specifies the shape of the input data. In most deep learning frameworks, like TensorFlow/Keras and PyTorch, we generally do not explicitly specify the batch size dimension when **defining** the input shape for layers in our model, including the dense (fully connected) layers. The batch dimension is typically assumed to be dynamic, allowing us to process different batch sizes without needing to redefine the model. When we define a dense layer in a Keras model, we can use the "input_shape" parameter to specify the shape of the input data, excluding the batch size. For example:

  ```
  from tensorflow.keras.models import Sequential
  from tensorflow.keras.layers import Dense

  model = Sequential()
  model.add(Dense(64, activation='relu',
              input_shape=(input_dimension,)))
  # ... more layers ...
  ```

  However, when we are actually feeding data into the model during training or inference, our data needs to have the appropriate batch dimension. This means the input data should be shaped with the batch size as the first dimension. In the example above, note that the single input dimension above is now fed with a 2D data array with the first dimension being the batch size.

  ```
  import numpy as np
  from tensorflow.keras.models import Sequential
  from tensorflow.keras.layers import Dense

  # Define the model
  input_dimension = 20  # Example input dimension
  model = Sequential()
  ```

```
model.add(Dense(64, activation='relu',
        input_shape=(input_dimension,)))
model.add(Dense(10, activation='softmax'))

# Compile the model
model.compile(optimizer='adam',
       loss='categorical_crossentropy',
        metrics=['accuracy'])

# Generate dummy data for demonstration
batch_size = 32  # Example batch size
# Create a batch of input data
        # (32 samples, each with 20 features)
input_data = np.random.random((batch_size,
            input_dimension))
# Create corresponding dummy labels
        # (32 samples, 10 classes for output)
labels = np.random.randint(10, size=(batch_size, 1))
labels = tf.keras.utils.to_categorical(labels,
            num_classes=10)

# Feed the data to the model
model.fit(input_data, labels, epochs=5,
          batch_size=batch_size)
```

- Output Layer Configuration: The configuration of the output layer depends on the specific problem (e.g., number of classes in classification tasks). If it is a binary classification, the number of units equals one. For categorical outputs, the number of units will be the number of categories.
  In particular, for multi-class classification output, the activation function will be "softmax." It turns the output into a probability distribution over the classes, where the output of each neuron corresponds to the probability that the input belongs to the respective class.

```
from keras.models import Sequential
from keras.layers import Dense

# Number of output categories
num_categories = 10

# Create a Sequential model
model = Sequential()

# Add hidden layers...

# Adding the output layer for categorical outputs
model.add(Dense(units=num_categories,
            activation='softmax'))
```

Often, we need to add a BatchNormalization layer after the dense layer and before the output layer, so we do not specify all the parameters for the dense layer in a single line.

```
from keras.layers import Dense, BatchNormalization,
            Activation

model.add(Dense(units=64))
model.add(BatchNormalization())
model.add(Activation('relu'))
```

4.2 Normalization Layers

The normalization technique normalizes the inputs of each mini batch to have a mean of zero and a variance of one. It is often used to stabilize and accelerate the training of deep neural networks. Keras provides two types of normalization:

1. Batch Normalization: These layers apply a transformation that maintains the mean output close to zero and the output standard deviation close to one. The normalization is used to reduce internal covariate shift. This refers to the change in the distribution of network activations due to the change in network parameters during training which can slow down the training process since layers continuously need to adapt to new distributions. This is why Batch Normalization is typically applied after a layer, but before its activation function.
2. Layer Normalization is another technique used in neural networks, similar in purpose to Batch Normalization but with key differences in its approach and applications. It was introduced in a 2016 paper by Jimmy Lei Ba, Jamie Ryan Kiros, and Geoffrey Hinton titled "Layer Normalization." Unlike Batch Normalization, which normalizes across the batch dimension, Layer Normalization normalizes across the features. In other words, for a given layer, Layer Normalization computes the mean and variance used for normalization from all of the summed inputs to the neurons in that layer.
   Layer Normalization does not rely on the batch dimension; it works well with different batch sizes and is particularly effective in tasks where the batch size is small or varies. Layer Normalization can be applied similarly to Batch Normalization, typically after the linear transformation and before the activation function:

```
model.add(Dense(64))
model.add(LayerNormalization())
model.add(Activation('relu'))
```

4.3 Dropout Layers

The dropout layers randomly set a fraction of input units to zero at each update during training time, which helps prevent overfitting. The key parameter for a dropout layer is the dropout rate, which specifies the fraction of the input units to be dropped during training. For example, rate = 0.2 means approximately 20% of the input units are set to 0 at each update during the training phase. The choice of the dropout rate is crucial: a rate that is too high may lead to underfitting, while a rate that's too low might not effectively prevent overfitting.

It is commonly used in fully connected (dense) layers of a network and is typically applied to the outputs of intermediate layers, but it can be used after any layer except the input layer.

While dropout can be used in convolutional layers, it is less common. Other regularization techniques, like data augmentation and batch normalization, are often preferred in convolutional neural networks (CNNs).

During testing or inference, usually a dropout is not applied. The network uses all its units, and the weights are scaled appropriately based on the dropout rate used during training. This scaling is typically handled automatically by frameworks like Keras.

```
from keras.models import Sequential
from keras.layers import Dense, Dropout

model = Sequential()
model.add(Dense(64, activation='relu'))
model.add(Dropout(0.5))  # Applying 50% dropout
model.add(Dense(10, activation='softmax'))
```

4.3.1 Flattening Layers

If we want to use a custom output layer to classify specific categories, we will need to add a flatten layer to the end of the hidden layers and a custom output layer. In the MNIST handwritten example, we flatten the input layer explicitly using the reshape method:

`xTrainFlattened = xTrain.reshape(len(xTrain),784)`

Using a flatten layer does exactly the same transformation; it simply reshapes a multidimensional vector to a 1D vector as illustrated in Figure 4-1.

4.3.2 Pooling Layers

MaxPooling and Average Pooling are two operations commonly used in the context of convolutional neural networks (CNNs). They are types of pooling layers that reduce the spatial dimensions (i.e., width and height) of the input feature maps, effectively downsampling the input. This reduction helps to decrease the

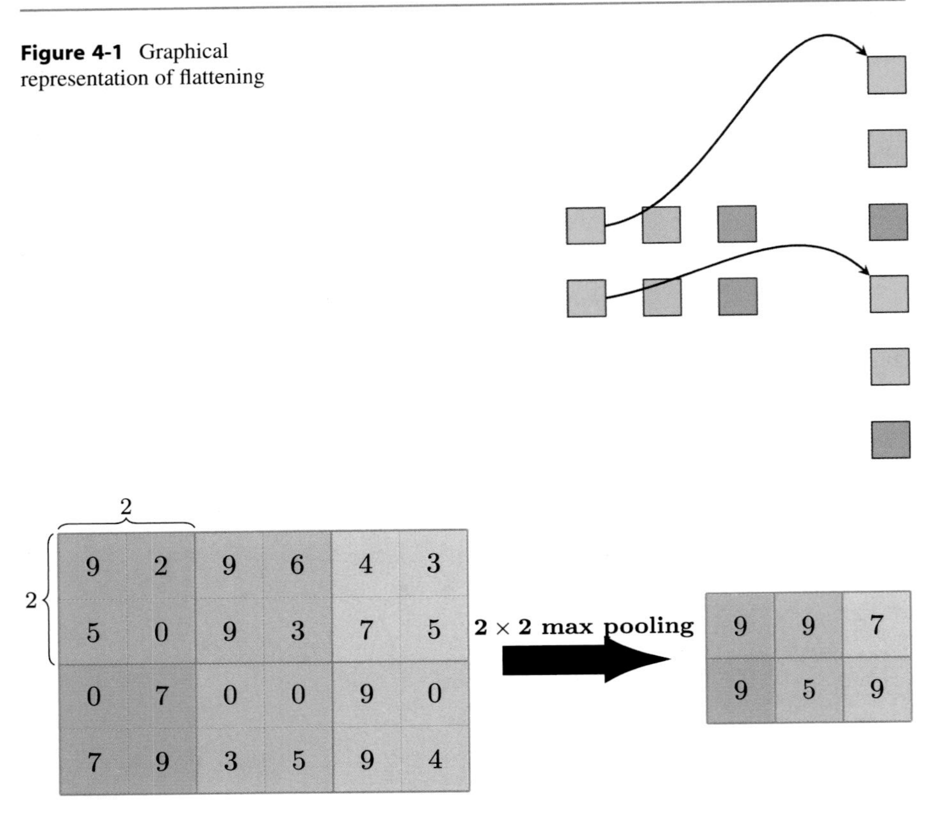

Figure 4-1 Graphical representation of flattening

Figure 4-2 Diagram showing MaxPooling2D with pool size=(2,2) and strides=(2,2)

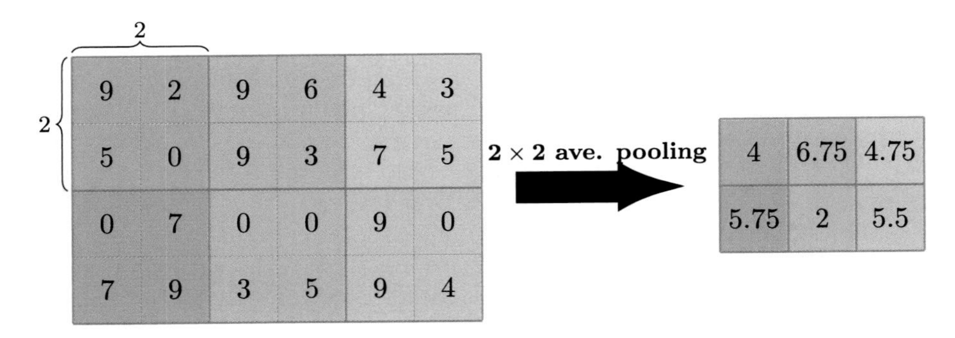

Figure 4-3 Diagram showing AveragePooling2D with pool size=(2,2) and strides=(2,2)

computational load, control overfitting by providing an abstracted form of the representation, and improve the network's ability to extract dominant features.

MaxPooling operates on each feature map independently. This process involves sliding a window (of a specified size and stride) over the input and outputting the maximum value within the window at each position. This process emphasizes the most pronounced features in each region of the feature map.

```
from keras.layers import MaxPooling2D

max_pool = MaxPooling2D(pool_size=(2, 2), strides=(2, 2))
```

Refer to Figure 4-2. In this example, pool size=(2, 2) means that the MaxPooling operation is applied over 2 × 2 windows, and strides=(2, 2) means that the window is moved 2 pixels across and 2 pixels down for each operation. This will effectively reduce the spatial dimensions of the feature map by a factor of 2.

Average Pooling also operates on each feature map independently. Similar to MaxPooling, it uses a window of a specified size and stride, but instead of taking the maximum value, it computes the average of the values in the window as shown in Figure 4-3.

```
from keras.layers import AveragePooling2D

# Example of an AveragePooling layer in Keras
average_pool = AveragePooling2D(pool_size=(2, 2),
    strides=(2, 2))
```

Here, the term "pool size=(2, 2)" indicates that the Average Pooling is applied over 2 × 2 windows, and strides=(2, 2) moves the window 2 pixels over and 2 pixels down. This reduces the spatial dimensions like MaxPooling, but it averages the values instead of taking the maximum.

MaxPooling is more commonly used because it generally performs better, focusing on the most salient features. It's especially effective in scenarios where the background of the input data is relatively uniform or less important.

Average Pooling can be more beneficial when we need to preserve background information or when the importance is more uniformly distributed across the feature map.

Both types of pooling help to make the representation approximately invariant to small translations, a desirable property in many vision tasks. The choice between them often depends on the specific requirements of the task and empirical performance.

4.3.3 Convolutional Layers

We will almost certainly encounter convolutional layers when dealing with image processing as they are used extensively for upsampling and downsampling images.

Convolutional layers are the core building blocks of CNNs. They perform a mathematical operation called convolution, which involves sliding a filter (or kernel) over the input data (like an image). As the filter moves across the input, it performs element-wise multiplication with the part of the input it covers and sums up these products to produce a feature map. This process extracts important features from the input, such as edges, textures, or specific shapes.

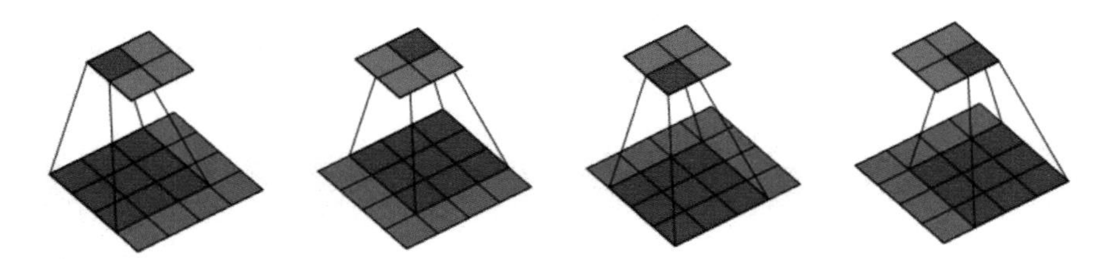

Figure 4-4 Convolving a 3 × 3 kernel over a 4 × 4 input using unit strides with no padding [1, p. 68]

Refer to Figure 4-4. In this example, the 3 × 3 kernel is shown in dark blue. With a unit stride, the model takes the values from the input image shown in dark blue and performs mathematical convolution with the kernel. At each position, an element-wise multiplication is performed between the values in the kernel and the corresponding values in the image it covers. The results of these multiplications are then summed up to get a single number. This sum is the output for the current position of the kernel. The operation results in an output of size (2,2) shown in green.

The weights of the kernels are learned during the training process. The network adjusts these weights to minimize the difference between its predictions and the actual data.

Stride refers to the number of units the filter moves across the input matrix. With a stride of one, the filter moves one unit at a time. This results in a detailed feature map, capturing more information. With a stride of two, the filter moves two units each time. This leads to a smaller feature map as it skips over more of the input. Strides greater than one are used for *downsampling* the input.

Padding for a convolutional layer is the process of adding extra pixels around the edge of the input. The most common types of padding are padding="valid" (no padding) and padding="same".

With no padding (valid padding), the size of the feature map is reduced as the filter cannot move beyond the edge of the input. padding="same" padding adds zeros around the input so that the output feature map has the same dimensions as the input. Refer to Figure 4-5. In this example, we introduce padding to enlarge the size of the 5 × 5 input in blue to an output size of 6 × 6 shown in green.

Declaring a convolutional layer in Keras is straightforward. The convolutional layer we will most commonly encounter is the Conv2D layer, which is typically used for processing images.

```
from keras.layers import Conv2D

conv_layer = Conv2D(filters, kernel_size, strides=(1, 1),
padding='valid', activation='relu', input_shape)
```

Filters The number of filters (kernels) in the convolutional layer. Each filter extracts different features from the input. In our previous example, we use a single

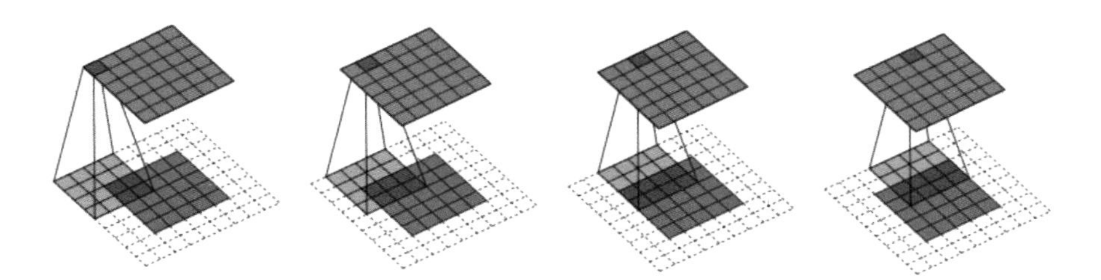

Figure 4-5 Convolving a 4 × 4 kernel over a 5 × 5 input padded with a 2 × 2 border of zeros using unit strides [1, p. 14]

filter, but there is no reason for that. Multiple filters are often use in many cases on the image.

Kernel Size The size of the filter. Common choices include (3, 3) or (5, 5). This parameter can be an integer in which case a square kernel is implied or a tuple of two integers to explicitly specify the x and y filter dimension.

4.3.4 CNN As an Input Layer

When used as an input layer in a neural network, the first convolutional layer typically processes the raw input data.

This layer must be configured with the shape of the input data (e.g., image dimensions and color channels), for example, an input shape of (28, 28, 1) for grayscale images of size 28 × 28 pixels or (224, 224, 3) for color images of size 224 × 224 pixels with three color channels (RGB). Note that the arrangement of dimensions for an RGB image can be specified in Keras using channel_first or channel_last format so that a (224, 224, 3) image is still valid as a (3, 224, 224) tensor if the parameter data_format= "channel_first" is specified. If a grayscale image is used, then the number of channels is one instead of three. An example of using the CNN as the first layer with the channel_first option is shown below:

```
from keras.models import Sequential
from keras.layers import Conv2D

model = Sequential()
model.add(Conv2D(32, kernel_size=(3, 3), activation='relu',
input_shape=(1, 28, 28), data_format='channels_first'))
```

The pixels in an RGB image is usually normalized to tensor with a range of [0,1] or [−1,1] before feeding into a CNN input layer. If the original pixel values are in the range of [0,255], one common approach is to scale these values down to a range between 0 and 1 by dividing all pixel values by 255, for example:

$$normalized\ image = \frac{image}{255.0}$$

Another common technique involves subtracting the mean and dividing by the standard deviation of the pixel values, either globally or per channel. This standardizes the pixel values to have a mean of zero and a standard deviation of one by subtracting the mean and dividing by the standard deviation:

$$normalized\ image = \frac{(image - mean)}{\sigma}$$

The mean and standard deviation can be computed over the entire dataset.

If we are using a pretrained model, it is important to normalize the images in the same way the model was originally trained. For example, models trained on the ImageNet dataset often use specific mean and standard deviation values for normalization. If unnormalized images are fed into a pretrained model, bizarre results can be seen with color shifts or patches of wrong colors appearing in the output.

Normalizing the input values helps accelerate training and improve performance by stabilizing the distribution of values within the network.

4.3.5 Multiple CNN Layers

Although we have shown a single CNN layer for illustration, in practice a series of CNN layers of different kernel sizes, strides, padding, etc., are used to upsample and downsample an image. Instead of using the image at its original size, the image is first "upsampled" via the use of nearest neighbor, bilinear interpolation, or transposed convolution to increase the effective resolution of the image. Toward the end of the CNN layer chain, the upsampled image is then downsampled to the output size.

The initial layers of a CNN typically learn to recognize simple patterns, like edges and basic textures. As we progress deeper into the network, later layers learn to recognize more complex features, like shapes or specific objects, by combining the simpler features extracted in earlier layers. This is why when we select a layer to use in a pretrained network, such as VGG-16, as shown in Figure 4-6, we can choose to use a layer in an earlier layer such as "conv2_1" or a deeper block to extract more abstract feature such as "conv4_3." One common problem with using CNN layers is the "checkerboard" issue, often referred to as "checkerboard artifacts, particularly in tasks involving image generation like autoencoders, generative adversarial networks (GANs), and super-resolution. These artifacts manifest as a checkerboard pattern in the generated images as shown in Figure 4-7. The issue primarily arises due

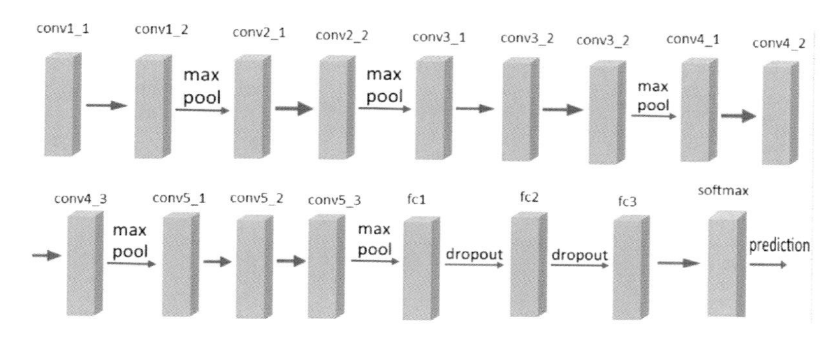

Figure 4-6 VGG-16 network block diagram [2, p. 3]

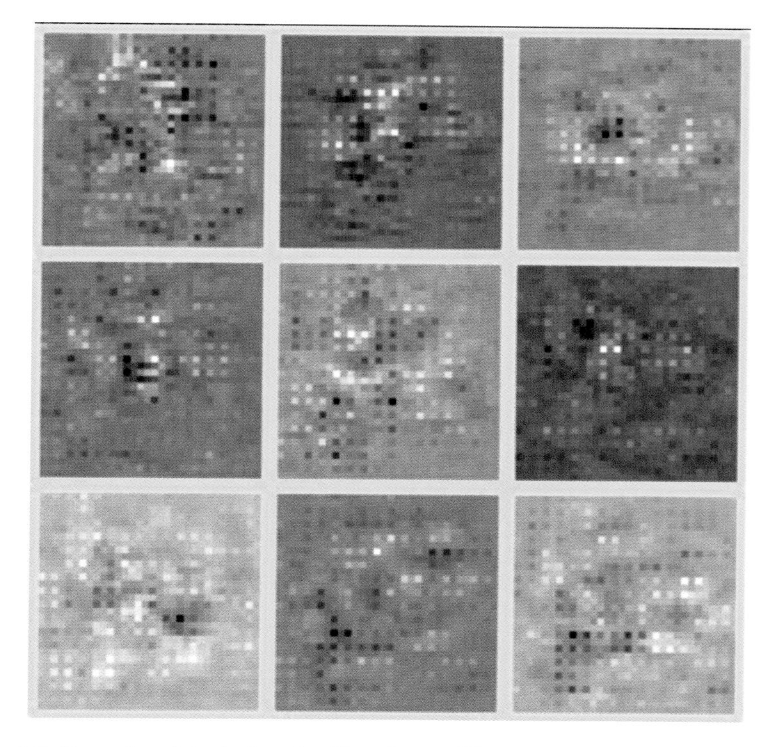

Figure 4-7 Examples of checkerboard artifacts from a CNN layer [3]

to the use of strided convolutions or transposed convolutions (sometimes called deconvolutions) for upsampling.

The checkerboard problem is caused when using strided or transposed convolutions for upsampling; the overlap in the convolution operation can be uneven. This uneven overlap can lead to certain pixels being updated more frequently than others, creating a visible grid-like pattern in the output.

The solution to this is to use a different non-overlapping stride sizes, change the kernel size, or alternative upsampling techniques. Often, we will find the researcher adding a convolutional layer to smear out the checkerboard effect with bilinear interpolation.

The GAN project included in this book also suffers from this problem, and the solution was to generate a more diverse set of data, including augmented cartoon images and longer training time.

4.3.6 Embedding Layers

An embedding layer is a specialized layer in neural networks, used primarily in the field of natural language processing (NLP), but also applicable in other areas where data can be converted into discrete tokens. For example, the BachBot application in this book uses an embedding layer.

The main function of an embedding layer is to convert these tokens (like words in text) into dense vectors of fixed size, which are more meaningful and suitable for performing various machine learning tasks, although in the case of BachBot, embedding the input did not produce a significant improvement in the results.

In text processing, words or phrases are typically represented as discrete tokens or integers. Each unique word in our vocabulary is assigned a unique integer ID.

The embedding layer transforms these integer tokens into dense vectors of a specified size. This size is a hyperparameter (that is chosen by us) and represents the dimensions of the embedding space.

Unlike one-hot encoded vectors which are high-dimensional and sparse, embedding vectors are low-dimensional and dense, containing real-valued numbers.

The vectors obtained from an embedding layer capture more information about words, including semantic meaning and contextual relationships. During training, these vectors are learned and adjusted to reduce the model's prediction error, making the embeddings contextual to the specific task.

Often, pretrained word embeddings, such as Word2Vec and Glove, are used in the embedding layer. These embeddings are trained on large datasets and capture a vast amount of semantic information.

Here is a simple example of how to use an embedding layer in Keras:

```
from keras.models import Sequential
from keras.layers import Embedding

model = Sequential()
model.add(Embedding(input_dim=vocab_size,
output_dim=embedding_dim, input_length=max_length))
```

4.3.7 Residual Layers

The key idea behind residual layers is the introduction of skip connections, also called shortcut connections or residual connections. These connections allow the input to a layer or a set of layers to be added to its output. This is typically done by adding the output of a convolutional block to its input so that the output of the block is the sum of the two. For instance, in a basic residual block, if the input is x and the output of the convolutional layers is $F(x)$, the final output of the block will be $F(x) + x$.

These skip connections help in easing the flow of gradients during backpropagation because they provide an alternative pathway for the gradient. This architecture alleviates the vanishing gradient problem and allows for training much deeper networks.

The skip connections often perform identity mapping, where the input is passed through unchanged to the output. When the input and output dimensions differ, a linear projection might be applied to match the dimensions.

```
from keras.layers import Input, Conv2D,
   BatchNormalization, Add
from keras.models import Model

# Input tensor
input_tensor = Input(shape=(256, 256, 3))

# First convolutional layer
conv1 = Conv2D(64, (3, 3), activation='relu',
             padding='same')(input_tensor)
conv1 = BatchNormalization()(conv1)

# Second convolutional layer
conv2 = Conv2D(64, (3, 3), activation='relu',
             padding='same')(conv1)
conv2 = BatchNormalization()(conv2)

# Skip Connection (identity mapping)
skip_connection = Add()([conv2, input_tensor])

# Creating the model
model = Model(inputs=input_tensor,
   outputs=skip_connection)
```

4.3.8 Recurrent Layers

An RNN layer is designed to deal with sequential and temporal problems, such as language translation, natural language processing (NLP), music generation, and image captioning.

Let's take an idiom, such as "Easier said than done," which is taken to mean not as easy as it appears to be. In order for the idiom to make sense, it needs to be expressed in that specific order. As a result, recurrent networks need to account for the position of each word in the idiom and use that information to predict the next word in the sequence.

A recurrent layer is a sweeping term referring to a group of specialized layers in Keras. There are too many variants to describe in detail here, so it is best to refer to the Keras manual for more specific details, but here is a brief description of what each type is used for:

1. LSTM Layer: Ideal for capturing long-term dependencies in sequential data. Commonly used in time series forecasting, natural language processing, and sequence prediction tasks. The BachBot project implementation in this book uses LSTM layers to capture music sequences.
2. GRU Layer: Similar to LSTM, but with a simpler architecture. Used for tasks like text generation, speech recognition, and time series analysis. Requires fewer parameters than LSTM, making it more efficient while still capturing long-term dependencies effectively.
3. SimpleRNN Layer: The most basic form of RNN, suitable for sequences where short-term context is sufficient. Useful in simpler sequence tasks.
4. TimeDistributed Layer: Applies a specified layer to each time step of a sequence independently. Commonly used with CNN layers in sequence-to-sequence tasks, like video processing or time series classification.
5. Bidirectional Layer: Wraps around another RNN layer (like LSTM or GRU) to process the sequence in both forward and backward directions. Widely used in NLP tasks like sentiment analysis and machine translation.
6. ConvLSTM1D, ConvLSTM2D, ConvLSTM3D: Used for a sequence of images or videos.

A distinguishing characteristic of recurrent networks is that they share parameters across each layer of the network. While feedforward networks have different weights across each node, recurrent neural networks share the same weight parameter within each layer of the network. These weights are still adjusted in through the processes of backpropagation through time and gradient descent to facilitate learning.

The backpropagation through time (BPTT) algorithm is slightly different from traditional backpropagation as it is specific to sequence data. The principles of BPTT are the same as traditional backpropagation, where the model trains itself by calculating errors from its output layer to its input layer. BPTT differs from the traditional approach in that BPTT sums errors at each time step, whereas feedforward networks do not need to sum errors as they do not share parameters across each layer.

Through this process, RNNs tend to run into two problems, known as exploding gradients and vanishing gradients. These issues are defined by the size of the gradient, which is the slope of the loss function along the error curve. When the gradient is too small, it continues to become smaller, updating the weight parameters until they become insignificant—that is, zero. When that occurs, the algorithm is no longer learning. Exploding gradients occur when the gradient is too large, creating an unstable model. In this case, the model weights will grow too large, and they will eventually be represented as NaN. One solution to these issues is to reduce the number of hidden layers within the neural network, eliminating some of the complexity of the RNN model.

We will look at an example of an LSTM layer:

```
from keras.models import Sequential
from keras.layers import LSTM

# Define the model
model = Sequential()

# Add an LSTM layer
# 'units' refers to the number of neurons
# in the LSTM layer
# 'input_shape' should match the shape of our
# training data
model.add(LSTM(units=50, activation='tanh',
         recurrent_activation='sigmoid',
input_shape=(time steps, features)))

# model.add(Dense(1))  # Example for a regression problem
```

It is important to understand what the parameters mean as they can be rather confusing with RNN. Refer to Figure 4-8. The subscript $t_1, t, t+1$ refers to the time step. The X_{t-1}, X_t, X_{t+1} are the input tensors at each time step; y_{t-1}, y_t, y_{t+1} are the outputs. The h_{t-1}, h_t, h_{t+1} are the outputs of the hidden states which are used in conjunction with the inputs Xs to get Ys. The W is the common weight tensor of the hidden states.

Each hidden state h_i has a number of units of hidden cells, four in this case, and this is the hyperparameter **units** needed for defining the LSTM layer.

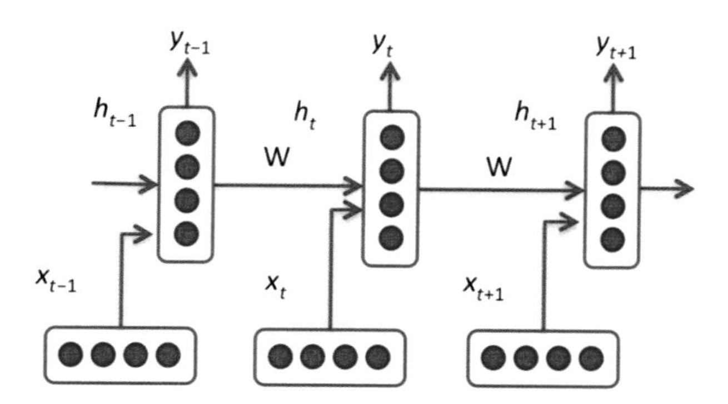

Figure 4-8 An RNN-LSTM network

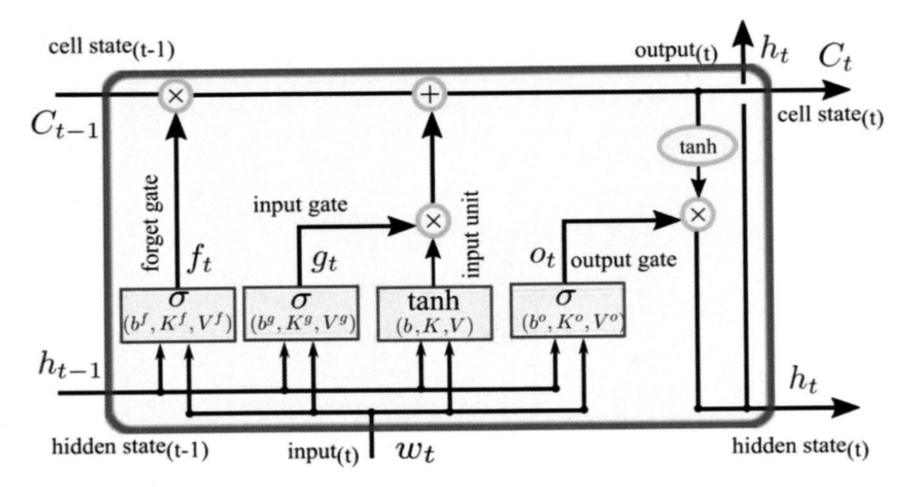

Figure 4-9 The block diagram of an LSTM cell

There are two activation functions specified in the example, activation='tanh' and recurrent_activation='sigmoid'. Unfortunately (because it is complex) to understand these terms, we need to examine the configuration of an LSTM unit.

Refer to Figure 4-9. The hidden state h_t is referred to as the actual output of the LSTM cell for each time step. It is derived from the cell state but is not the same as the cell state. The hidden state is calculated using the cell state and the output of the output gate. It's typically passed to subsequent layers in the network or used as the final output in sequence processing tasks.

The cell state c_t is the internal memory of the LSTM cell. It carries information across time steps and is updated at each time step based on the previous cell state, the current input, and the outputs of the forget and input gates. The cell state is used as an internal mechanism that allows the LSTM to maintain a memory over time, so we do not refer to it as an output of the LSTM cell.

At each time step, the LSTM cell takes the previous cell state c_{t-1} and the previous hidden state h_{t-1}, along with the current input x_t, to calculate the new cell state c_t and the new hidden state h_t. The new hidden state h_t is then propagated forward in two directions: horizontally to the next time step $t + 1$ in the sequence and vertically to the next layer if the LSTM is part of a stacked LSTM architecture.

Notice that there are two activation functions for an LSTM cell, and these are the parameters required when setting up an LSTM layer in Keras.

4.3.9 Activation Function

This parameter refers to the activation function used for the LSTM cell's output. The activation function is applied to the cell state before it is outputted. This is akin to the activation function in a traditional feedforward neural network layer. Common choices for the activation function are nonlinear functions, like tanh or ReLU. tanh is the most common choice in LSTMs, as it outputs values in a range between −1 and 1, which is useful for normalizing the output of the LSTM.

4.3.10 Recurrent Activation

The recurrent_activation argument applies to the input, forget, and output gates. The recurrent activation function is used to calculate the state of these gates and thus regulates the flow of information through the cell.

The common choice for the recurrent activation function is the sigmoid function. The sigmoid function outputs values between zero and one, making it suitable for gating purposes (like deciding how much of the previous state to keep or how much of the new state to write).

The parameters b, V, W are respectively biases, input weights, and recurrent weights. In the LSTM architecture, the forget gate uses the output of the previous hidden state cell to control the cell state C_t to remove irrelevant information. On the other hand, the input gate and input unit add new information to C_t from the current input.

4.3.11 Other Layers

There are other types of specialized layers which are out of scope for this book. In particular, the attention layer is used extensively for large language models (LLMs). These layers allow the model to focus on specific parts of the input sequentially, rather than using the entire input at once. They are a key component in transformer models, which are used in various NLP tasks.

Part II

Implementation Examples

5 The Training Process

As much as we would like to throw the data into a machine learning model and get instant insights into the data, this is not the way it will work. In most cases, the data have to be scrubbed, transformed into model-readable format, and augmented. In addition, we also need to select and design an appropriate machine learning algorithm for the task.

Data preprocessing is a relatively standard procedure in most cases. Selecting the right algorithm can be overwhelming, particularly when the model neural network names are highly complex. For example, we will show examples of convoluted, recurrent, long short-term memory, and generative adversarial neural networks in this book. The naming convention, although confusing at first, will be helpful as a memory aid once we have understood the intended functionalities as the name often refers to the topology of the network itself. Refer to Figure 5-1. The diagram shows a typical end-to-end machine learning process. The data loading and productionizing of the model stages are the same as for other IT projects, although we will discuss some data types in Python which would make the data loading easier for machine learning. The middle three stages are specific to machine learning, which we will discuss in detail below.

5.1 Data Loading

In a commercial environment, it is most likely that the data for the project would be stored in a relational database and would require coding using SQL for retrieval. For learning and smaller datasets stored in a CSV file or online via a URL, we can use NumPy or Pandas to read the data in directly.

P. Hua, *Neural Networks with TensorFlow and Keras*,
https://doi.org/10.1007/979-8-8688-1020-6_5

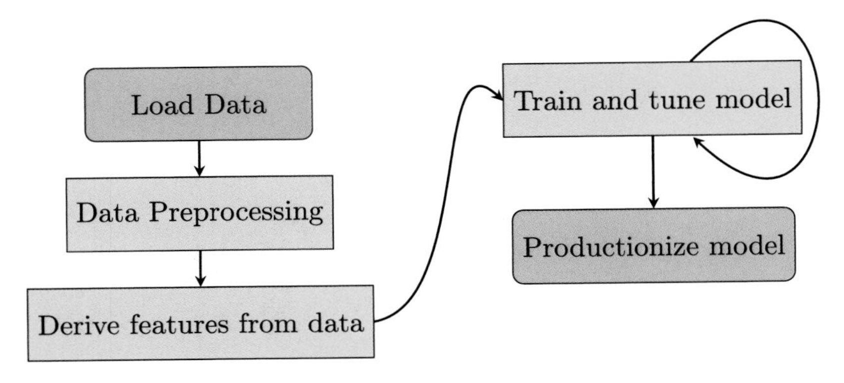

Figure 5-1 End-to-end machine learning process

For this demonstration, the COVID-19 database from Kaggle will be used. The COVID-19 database can be downloaded from Kaggle using the URL to the current local directory:

https://www.kaggle.com/datasets/meirnizri/covid19-dataset

The dataset is approximately 5MB in size, in CSV format, making it manageable for processing. We will be using this dataset as a project later on, so we would just be loading it for now.

Using Pandas to load data into a pandas.DataFrame is probably the most flexible way, allowing us to summarize the data immediately. If using Google Colab, upload the file by dragging it into a Colab directory. As an example, use the procedure below to upload the file `/content/sample_data/Covid Data.csv` to Colab.

The first line in the file contains the column headings, so we assign `data.columns = iloc[0]`; otherwise, we would have to assign the column names explicitly as

`pandas.DataFrame(data,columns=['USMER','MEDICAL_UNIT',...]`

```
import pandas

filename = "/content/sample_data/Covid Data.csv"
data = pandas.read_csv(filename)
pandas.columns = data.iloc[0] # use the first row as names
data = data[1:]
print(data.shape)
print(data.head(5))
```

```
(1048574, 21)
   USMER  MEDICAL_UNIT  SEX  PATIENT_TYPE   DATE_DIED ...
1      2             1    2             1  03/06/2020  ...
2      2             1    2             2  09/06/2020  ...
3      2             1    1             1  12/06/2020  ...
4      2             1    2             1  21/06/2020  ...
5      2             1    1             2  9999-99-99  ...

  AGE  PREGNANT  DIABETES  ...  ASTHMA  INMSUPR
```

```
1   72          97          2  ...       2        2 ...
2   55          97          1  ...       2        2 ...
3   53           2          2  ...       2        2 ...
4   68          97          1  ...       2        2 ...
5   40           2          2  ...       2        2 ...

    CARDIOVASCULAR  OBESITY  RENAL_CHRONIC  TOBACCO ...
1                2        1              1        2  ...
2                2        2              2        2  ...
3                2        2              2        2  ...
4                2        2              2        2  ...
5                2        2              2        2  ...

[5 rows x 21 columns]
```

```
data.describe()
USMER  MEDICAL_UNIT             SEX  PATIENT_TYPE ...
count  77064.000000  77064.000000  77064.000000 ...
mean       1.486271      3.740956      1.567243 ...
std        0.499815      0.456079      0.495461 ...
min        1.000000      1.000000      1.000000 ...
25%        1.000000      3.000000      1.000000 ...
50%        1.000000      4.000000      2.000000 ...
75%        2.000000      4.000000      2.000000 ...
max        2.000000      4.000000      2.000000 ...
```

Pandas has several useful functions to give a high-level view of the data. `data.describe()` gives the overall statistics for the data. It is not particularly useful in this case since the data is mostly categorical but would prove beneficial for other numerical datasets. data.info() tells us if the columns have null values.

```
<class 'pandas.core.frame.DataFrame'>
RangeIndex: 77064 entries, 1 to 77064
Data columns (total 21 columns):
#   Column                Non-Null Count  Dtype
---  ------                --------------  -----
0   USMER                 77064 non-null  int64
1   MEDICAL_UNIT          77064 non-null  int64
2   SEX                   77064 non-null  int64
3   PATIENT_TYPE          77064 non-null  int64
4   DATE_DIED             77064 non-null  object
5   INTUBED               77064 non-null  int64
6   PNEUMONIA             77064 non-null  int64
7   AGE                   77064 non-null  int64
8   PREGNANT              77064 non-null  int64
9   DIABETES              77064 non-null  int64
10  COPD                  77064 non-null  int64
11  ASTHMA                77064 non-null  int64
12  INMSUPR               77064 non-null  int64
13  HIPERTENSION          77064 non-null  int64
14  OTHER_DISEASE         77064 non-null  int64
15  CARDIOVASCULAR        77064 non-null  int64
16  OBESITY               77063 non-null  float64
17  RENAL_CHRONIC         77063 non-null  float64
```

```
18  TOBACCO                  77063 non-null  float64
19  CLASIFFICATION_FINAL     77063 non-null  float64
20  ICU                      77063 non-null  float64
dtypes: float64(5), int64(15), object(1)
memory usage: 12.3+ MB
None
```

We can see from the summary that columns 16 to 20 have missing data. A more accurate way to count the total number of missing data is by using `data.isnull().sum().sum()`. This returns five missing data points as expected. Once we know there are null values, we can use

`nullValues = data[data['OBESITY'].isnull()]`

to return the rows with missing values for a particular column. The `isnull()` function works for NaN data as well, so there is no need to use `isna()` separately.

It is possible to use a pandas DataFrame anywhere a NumPy array is used if we have a data type of real or integer. This works because the pandas.DataFrame class supports the `__array__` protocol, and TensorFlow's `tf.convert_to_tensor` function accepts objects that support the protocol. Alternatively, we can convert the dataframe directly to NumPy array or tensor using the following commands:

```
numpy.array(data)
```

or

```
tf.convert_to_tensor(data)
```

In many cases, we need to transform or rescale the input data before using the model. This is normally done by adding column(s) to the original dataset using

```
DataFrame.insert(loc=location, column="column name")
```

and applying the necessary logic to create the new data. For example, in the COVID-19 dataset, we may want to create a column to specify if the patient has existing medical conditions as

```
data.insert(column="PreExisting Condition")
```

Loading the data via a URL is also straightforward using Pandas, and the code is the same except that instead of passing the path of the local file, we would pass the URL to the `read_csv()` method. As before though, if we use Colab, then the file needs to be uploaded onto Google Drive before it can be read.

5.1.1 Loading Images

In image classification tasks, it's common to train models using hundreds of thousands of images with convolutional neural networks. Fortunately, Keras offers several utilities to streamline the process of image loading and preprocessing. If the directory structure is

```
main_directory
\ class a
```

```
image 1.jpg
image 2.jpg
\ class b
image 3.jpg
image 4.jpg
...
```

then calling `tf.keras.utils.image_dataset_from_directory()` will generate a dataset from the files in the subdirectories and assign automatic labels to each group of images, so it will return tuples in the form of `images,labels`. The complete list of parameters is explained under:

https://keras.io/api/data_loading/image/

Most parameters are self-explanatory; however, the following ones require further explanation:

- Labels: There are three options. "inferred" labels are generated from the directory structure, for example, class a, class b, in conjunction with the parameter label mode. If the label mode is `int`, the labels will be encoded as integers to be used with `sparse_categorical_crossentropy` loss function. "categorical" label mode means the labels will be one-hot encoded for `categorical_crossentropy` loss. When there are only two classes, "binary" should be used with `binary_entropy` loss function.
- Batch size: A machine learning algorithm normally processes images in batches and updates the neuron weights after each batch size instead of a single image.
- Validation Split: If this is set to 1, then the subset parameter will need to be specified as below.
- Subset: One of "training," "validation," or "both." When subset="both", it returns a tuple of training and validation datasets.

The Keras utility also allows the loading of individual images using `tf.keras.utils.load_img()`. The point to note is that the utility returns a PIL image instance, and this needs to be converted to a NumPy array using `tf.keras.utils.img_to_array` as data for the input layer. This process is typically sufficient in most cases. However, when working with pretrained networks, additional preprocessing may be required to ensure the images are in the correct format. Since there is no standardization for this step, it is essential to consult the specific instructions for the input layer of each pretrained network.

5.2 Data Processing

After loading the data, it is necessary to prepare the data in a format that could be fed into the neural network. The main objective for the data preparation is to clean the data as much as possible so that they can be passed as features into the model. In some cases, particularly when we do not have enough data for the machine to train, some data augmentation is also necessary to improve the model's accuracy. There are many issues with data, but the common ones are dealt with below.

5.2.1 Splitting the Dataset: Training, Development, Test

Machine learning is an iterative process. Choosing the right model with the right set of parameters at the outset is impossible, so practically we have to try out different models and parameters and refine them until our goals are achieved. When we approach machine learning this way, it is standard practice to split the dataset into separate training, development, and test datasets as shown in Figure 5-2.

The idea is that we use the training dataset to experiment with different models and try out different parameters. The development dataset, also called the holdout or cross-validation dataset, is then used to compare the generalization performance of different models on unseen data. We do not use the training dataset for this purpose to avoid the risk of shortlisting the best overfitted models as discussed in the previous section. The development dataset is also used for fine-tuning hyperparameters of the models, such as the step size or the optimization algorithm. We do not use the training data to select model parameters as this will lead to suboptimal hyperparameters, giving bias toward models that will overfit.

The best model selected from the development set is then evaluated using the test data. If we have a smallish dataset, say less than 10,000 data points, then the split between the three training/dev/test datasets is normally in the ratio of 60%:20%:20%. Commercial projects with 1,000,000+ samples use much more data to train and less to tune and validate, so the ratio could even be in the order of 99%:0.5%:0.5% or even less for the development and test data.

To build a well-performing model, it is essential to train and test the data which came from the same distribution. This means, for instance, that we should be looking broadly at the same type and quality of data: clear pictures of cats taken in similar surrounding for the datasets. What we want to avoid in this example is to have the model trained on domestic cats and evaluated on blurry pictures of lynxes. In some cases, it may not be possible to have a big dataset for the model to train on. For these cases, there are tricks that we could use to augment the data, which will be discussed in Section 5.2.6.

5.2.2 Categorical Data

When data is represented as a list of categories or finite objects, for example, {apples, oranges, strawberries, bananas, grapes}, some algorithms, such as decision trees, can deal with the text directly, and there is no need for any data encoding. For many other algorithms, the data needs to be converted into a list of numerical IDs. It may be sufficient to assign a numeric ID to each item, for example, {apples=1,oranges=2,...}. However, integer encoding may lead the algorithm to imbue a spurious relationship between different categories because of numerical

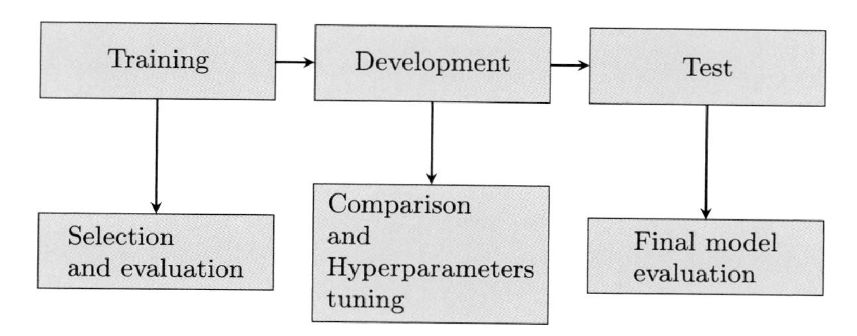

Figure 5-2 Splitting the data into training, development, and test datasets

ordering. Assuming natural ordering between the numerical categories may result in poor performance and/or unexpected results.

One-hot encoding is a binary encoding scheme where each item in the list of categories is assigned a value of one if true and zero if not. The unique list of categories stored in the rows is translated as columns in Pandas and assigns a corresponding one and zero using the method `get_dummies()` as in the following example:

```
import pandas

data = pandas.DataFrame(list(['apples','oranges',
'strawberries','bananas','grapes']), columns = ['fruits'])
converted = pandas.get_dummies(data.fruits, prefix='onehot')
print(data)
print(converted)
```

```
    fruits
0        apples
1       oranges
2  strawberries
3       bananas
4        grapes
onehot_ onehot_  onehot_ onehot_  onehot_
apples  bananas  grapes  oranges  strawberries
  1        0        0       0         0
  0        0        0       1         0
  0        0        0       0         1
  0        1        0       0         0
  0        0        1       0         0
```

The term "dummy" refers to the use of a statistical dummy variable in regression analysis. In statistics, a dummy variable acts like a switch in a regression equation, turning the parameter on and off by setting one or zero without the need to write multiple equations with different variables.

5.2.3 Preprocessing Images

Training a neural network to classify images normally means loading a custom dataset with image files in one of JPEG, BMP, or PNG format. The pixels then have to be converted into a NumPy array or tensor of type float and resized to match the size of the input layer of the neural network. Although not strictly necessary, the pixel values are normally rescaled to the range 0 to 1 or −1 to 1, which may speed up the training process. As an example, we will use the VGG-16 pretrained network in Keras to classify pictures of cats and dogs.

A number of pretrained networks are available in Keras as applications in two parts: model architecture and weights. The model architectures are already downloaded when we installed Keras; the weights are stored in large files which need to be downloaded when we instantiate a model. The reader can read up on the different models in Keras on the official website:

https://keras.io/api/applications/

As we will be using the VGG-16 network, let's discuss some of its properties which we need to know so that we can use it correctly.

The VGG-16 network is a convoluted neural network (CNN) which is designed for image recognition. It is used for object detection and image classification able to classify 1000 different classes of objects with 92% accuracy.

The list of classes can be found at

```
https://image-net.org/challenges/LSVRC/2014/browse-synsets
```

which includes classes of animals, such as cats, dogs, whales, etc.

The network was trained on color images of size 224 by 224 pixels with three color channels, so we need to convert our input images to this size. In addition to adjusting the image size, we also need to use the

```
keras.applications.preprocess_input()
```

method to convert the RGB colors for the input data to BGR, which is the color format used by VGG-16.

We do not need to do anything to the hidden layers other than to use the pretrained weights and tell the network to predict our image as one of the 1000 different classes.

In Code 5-1 below, the parameter `include_top=True` means we are using the complete network. It is possible just to use the pretrained network hidden layers and specify custom-sized input and output layers by setting this parameter to false. The pretrained network then can be repurposed and further trained on our own set of data.

Using a pretrained network this way is called *transfer learning*. It allows the model trained on one task to be repurposed for a related task with much quicker retraining. The variable `predictionLabels` is a list containing triplets in the format below. The first item is the label code, the second is the label text, and the number is the prediction probability.

```
<class 'list'> [('n02123045', 'tabby', 0.560446)]
```

```
# load model
vgg16 = keras.applications.vgg16.VGG16(include_top=True,
weights='imagenet')

trainingDs = tf.keras.utils.image_dataset_from_directory(
'/content/drive/MyDrive/dog vs cat/dataset/training_set/',
labels="inferred",label_mode="binary",image_size=(224,224),
batch_size=32)
fig = plt.figure(figsize=(10, 10))
for images, labels in trainingDs.take(1):
print(len(images),len(labels))
for i in range(16):
ax = plt.subplot(4, 4, i + 1)

# set the spacing between subplots
plt.subplots_adjust(left=0.1,bottom=0.1,
right=1.0,top=0.9,
wspace=0.4,hspace=0.4)

numpyImage = img_to_array(images[i])
numpyImage = np.expand_dims(numpyImage.copy(),axis=0)
x = preprocess_input(numpyImage)

prediction = vgg16.predict(x)
predictionLabels = decode_predictions(prediction)
plt.imshow(images[i].numpy().astype("uint8"))
aTitle = predictionLabels[0][0][1] + " prob " +
f'predictionLabels[0][0][2]*100:.0f' + '%'
plt.title(aTitle)
plt.axis("off")
plt.savefig('catsdogs.jpg')
```

Code 5-1 Using the VGG-16 pretrained network to classify cats and dogs

The network returns the list sorted by probability from high to low, so the first item in the list is the highest probability. This is why we can use `predictionLabels[0][0][1]` to get the label name for the most likely class of animal.

It is quite remarkable that we are able to use a sophisticated network to predict pictures of cats and dogs using a single line of code. Upon reviewing the model's results, it becomes evident that while effective, the model is not 100% perfect. Some misclassified pictures are clearly wrong to the viewer but perhaps forgivable if we view at the process for what it is—the detection of similar patterns in new pictures using weights which have been trained on a group of images.

5.2.4 Normalization and Standardization

It is often the case that the input variables have different scales. For example, a company's profit in dollars could be in billions, while its profit margin is expressed in percentage. Input variables may also have different units, such as cm, km, or miles.

When we model a problem with input of different scales, the weights for the model may also be large. After all, we are trying to approximate a function that fit the data points. The optimizer used in the backpropagation algorithm also works better when the scales are uniformed across the different variables. The goal of normalization is to change the values of numeric columns in the dataset to use a common scale, without distorting differences in the ranges of values without losing information. Normalization and standardization are crucial steps in data preprocessing, ensuring that each input feature contributes equally to the analysis.

Transformation can be done manually by using

```
sklearn.preprocessing
```

transformation or standardization methods on the input column(s), or we can use a Normalize layer in Keras to perform rescaling for us. If we transform data, it is important that the parameters are calculated for the training set and then used for the development and test datasets.

There are several transformation methods in sklearn, but two useful methods are `fit_transform()` and `transform()` for the MinMaxScaler and StandardScaler classes. StandardScaler normalizes data to a mean of 0 and a standard deviation of 1, making it suitable for algorithms that assume data is centered around zero. MinMaxScaler, on the other hand, scales data to a specified range, often [0, 1], which is useful in neural network applications.

The StandardScaler `fit_transform()` uses the following formula:

$$x_{transform} = \frac{x - \mu}{\sigma} \tag{5.1}$$

where μ is the average value of the dataset and σ is the standard deviation, whereas the MinMaxScaler will use

$$x_{transform} = \frac{x - x_{x\,min}}{x_{max} - x_{min}} \tag{5.2}$$

For example, we can see below that the transformed data using the standard scaler has mean zero and a unit standard deviation regardless of the original data, whereas the MinMaxScaler will scale differently and is more values dependent. However, the MinMaxScaler is useful for image processing when we want to compress the [0,255] color range down to [0,1].

```
from sklearn.preprocessing import MinMaxScaler
from sklearn.preprocessing import StandardScaler
import numpy as np

trainingData = np.array([1.0,2.0,3.0,4.0,5.0,6.0])
scaler = StandardScaler()
transformedData = scaler.fit_transform(trainingData.
        reshape(-1,1))
print(transformedData.mean())
print(transformedData.std())

mmScaler = MinMaxScaler()
mmTransformed = mmScaler.fit_transform(trainingData.
        reshape(-1,1))
print(mmTransformed)
print(mmTransformed.std())
```

```
-3.700743415417188e-17
1.0
[[0. ]
[0.2]
[0.4]
[0.6]
[0.8]
[1. ]]
0.34156502553198664
```

For convenience, Keras allows the scaler to work with Pandas DataFrame directly, so it is easier to perform transformation on the data columns directly.

```
scaler = StandardScaler()
dfScaled = pandas.DataFrame(scaler.
       fit_transform(data[['AGE']]))
```

5.2.5 Missing Data

Handling missing data is a common preprocessing challenge. Strategies include removing rows or columns with missing values or imputing these values based on the rest of the dataset. Missing data in columns or rows can be dealt with by removing the relevant row(s) or column(s) if they are not significant to the whole dataset, or we can look to use an algorithm in Keras/Pandas to impute missing values.

In the first instance, we can use a `dropna()` method in Pandas to delete any row or column in a DataFrame which has missing values. For example:

```
import pandas as pd
import numpy as np

missingData = 'col1':[1, 2,3],'col2':[3, np.NaN,5],
                    'col3':[6,7,np.NaN]
df = pd.DataFrame(data=missingData)
print(df)
noMissingRow= df.dropna(axis=0)
print('No missing row',noMissingRow)
noMissingCol= df.dropna(axis=1)
print('No missing col',noMissingCol)
```

```
  col1  col2  col3
0     1   3.0   6.0
1     2   NaN   7.0
2     3   5.0   Na
N
No missing row col1  col2  col3
0                 1   3.0   6.0

No missing col col1
```

```
0           1
1           2
2           3
```

Pandas also has a method `fillna()` to fill in missing data. Backfilling can be done for the dataframe or by rows and columns. The syntax for this method has several useful options, including filling missing data using data from the first available data before or after the missing column or row.

```
dataframe.fillna(value, method, axis, inplace, limit,
    downcast)
```

This returns a new dataframe if the inplace parameter is `false` and `None` if it is true. Some examples of `fillna()` are shown below:

```
import numpy as np
import pandas as pd

df1 = pd.DataFrame('Column1': [1.0,2.0,None],
'Column2': [3.0,None,5.0],
'Column3': [None,6.0,7.0])

df1.fillna(0)
df1.fillna(value='Column1': 0.1,'Column2': 0.2)
#original dataframe
print(df1,'original dataframe')
# forward fill using available number in the previous row
dff = df1.fillna(method='ffill',axis='rows')
print(dff,'forward fill')
# backward fill using available number in the row after
dfb= df1.fillna(method='bfill',axis='rows')
print(dfb,'backward fill')

 Column1  Column2  Column3
0      1.0      3.0      NaN
1      2.0      NaN      6.0
2      NaN      5.0      7.0 original dataframe
Column1  Column2  Column3
0      1.0      3.0      NaN
1      2.0      3.0      6.0
2      2.0      5.0      7.0 forward fill
Column1  Column2  Column3
0      1.0      3.0      6.0
1      2.0      5.0      6.0
2      NaN      5.0      7.0 backward fill
```

Using dropna() simplifies the dataset but can lead to loss of valuable information. fillna(), while preserving data, requires careful choice of imputation strategy to avoid introducing bias.

As well as Pandas, the scikit-learn library offers a convenient way to impute values by calling the `SimpleImpute` class. One of the most common interpolation techniques is mean imputation where we simply replace the missing values in each column with the column's mean or median value. The scikit-learn `SimpleImputer` method is more flexible as it offers more backfill option and can work on nparray data type as well as a dataframe.

```
import pandas as pd
import numpy as np
from sklearn.impute import SimpleImputer

missingData = 'col1':[1, 2,3,3,3,3],'col2':[3,
        np.NaN,5,4,4,4],
        'col3':[6,7,np.NaN,8,8,8]
df = pd.DataFrame(data=missingData)
print(df)
sim = SimpleImputer(missing_values=np.nan, strategy='mean')
simMedian = SimpleImputer(missing_values=np.nan,
        strategy='median')
imputedData = sim.fit_transform(df.values)
imputedMedian = simMedian.fit_transform(df.values)
print('imputed mean')
print(imputedData)
print('imputed median')
print(imputedMedian)
```

```
 col1  col2  col3
0     1   3.0   6.0
1     2   NaN   7.0
2     3   5.0   NaN
3     3   4.0   8.0
4     3   4.0   8.0
5     3   4.0   8.0
imputed mean
[[1.  3.  6. ]
[2.  4.  7. ]
[3.  5.  7.4]
[3.  4.  8. ]
[3.  4.  8. ]
[3.  4.  8. ]]
imputed median
[[1. 3. 6.]
[2. 4. 7.]
[3. 5. 8.]
[3. 4. 8.]
[3. 4. 8.]
[3. 4. 8.]]
```

5.2.6 Data Augmentation

Data augmentation is a powerful technique to enhance the size and quality of training datasets by introducing variations. This is especially important in fields like image processing, audio analysis, and natural language processing.

It is useful to create more input data with transformed data using augmentation. By applying augmentation, we can increase the model's capacity to generalize and make better predictions. Data augmentation can be used for many applications, including text, audio, and images. For images, we can perform geometric transformation such as scaling, rotating, flipping, cropping, kernel filtering (sharpening or blurring), or mixing images.

For audio, we can shift tone, balance, or speed or inject noise into an audio transcript to simulate dropout. We can even use an advanced library, such as Dolby noise reduction, to remove background noises.

For natural language processing, it is more difficult to augment data due to the grammatical structure of the text. Augmentation can be performed at character, word, or sentence level.

Some common techniques are used to create synthetic sentences, such as back translation, that is, we translate an English sentence into a foreign language and then back-translate the foreign sentence into English again to hopefully generate a different sentence from the original one. One commonly used and effective technique is synonym replacement via word embedding. The N-word embedding algorithm, for example, replaces N non-stopwords by pretrained synonyms.

Several libraries are available to use for this purpose, but they are divided into two categories: for non-contextual word embedding, models such as Glove and Word2Vec. The more advanced models, which use so-called `transformer` layers for learning contextual information, are BERT and RoBERTa from Google.

Implementing data augmentation in Keras is straightforward; we can either use the TensorFlow library to perform the necessary transformation, such as image rotation or word embedding, and use them as extra input data, or we can use Keras preprocessing layers for data augmentation, for example, `tf.keras.layers.RandomCrop` or `tf.keras.layers.RandomRotation`, and make the processing layers part of our model. Note that the data augmentation should only be active during training and not in production as we wish to supplement the input data to train the model to improve its score and not to create fake live data.

We will now look at some code snippets to augment images, sound, and sentences. In Keras, data augmentation can be easily implemented using preprocessing layers. Below is an example of how to resize and rotate images using these layers.

Example 1: The example below uses the Keras preprocessing layer to resize and randomly rotate images of roses for the input layer before sending them to the next layer in the network. With this option, preprocessing will happen on the device, synchronously with the rest of the model execution, so that it will maximize GPU acceleration. If we are training on a GPU, this is the best option for the normalization, image preprocessing, and data augmentation layers. The second option is to loop through each image and save them as new images as shown by the `augment()` function. An example of the output from the `augment()` function applied to a rose image is shown in Figure 5-3.

```
import tensorflow as tf
import pathlib
import matplotlib.pyplot as plt
import matplotlib.image as mpimg
from PIL import Image
from keras import layers
```

```
import numpy as np

def augment(oldImage):
image = oldImage.copy()
image = tf.image.random_crop(image, size=[128, 128, 3])
image = tf.image.random_flip_left_right(image)
return image

image = Image.open("/content/drive/MyDrive/Data/rose.jpg")
tensor = np.array(image)
sizeScaleRotate = tf.keras.Sequential([
layers.Resizing(128, 128),
layers.Rescaling(1./255),
layers.RandomRotation(np.pi)
])
imageAfter = sizeScaleRotate(tensor)

fig = plt.figure(figsize=(10, 7))
plt.subplot(1,3,1)
plt.imshow(image)
plt.subplot(1,3,2)
plt.imshow(imageAfter)

model = tf.keras.Sequential([
# Add the preprocessing layers
sizeScaleRotate,
# the rest of the model
layers.Dense(32, activation='relu')
])
# augment data separately
imageAfter2 = augment(tensor)
plt.subplot(1,3,3)
plt.imshow(imageAfter2)
```

Figure 5-3 Original image (left), resize and rotate (middle), crop and flip left-right (right)

5.3 Tuning Our Network

The machine learning process is highly iterative. Given that it is relatively easy to code up a simple neural network in Python, it is often the case that we would code up, evaluate, and recode until our objectives are met. An important part of this iterative process is the ability to evaluate the current model and the knowledge to improve it if needed.

There are several parameters to choose when we are designing a model, namely, the number of units, layers, choice of loss functions, optimization procedure, number of epochs, the activation function, and the learning rate of the model. There are two types of parameters in machine learning: model and algorithm hyperparameters.

- Model Hyperparameters: These are parameters for our model, such as the number of nodes and layers.
- Algorithm Hyperparameters: These are usually the parameters for the backpropagation algorithm, such as the learning rate.

Specialized networks will include additional parameters to select, but the parameters mentioned are the basic parameters in most deep neural network (DNN) models.

A common discussion on the network performance is the trade-off between the level of bias and variance. This is just a technical term for the degree of overfitting (high variance) versus underfitting (high bias). As mentioned before, a high variance network will have a low error rate with the training dataset but a high error rate with the validation dataset because it does not generalize well. Underfitting or high bias means the network is not sufficiently complex to deal with the problem.

A network with high bias and variance will have a large error both in the training and validation datasets. Of course, we want a low bias, low variance network if possible, but, usually, reducing the error for one means increasing the error for the other, hence the commonly known "bias-variance trade-off" in machine learning.

It is a fallacy to assume that a highly complex network will have high variance (i.e., prone to overfitting). A large language model, such as GPT-4, can have billions of parameters but generalize exceptionally well when compared to lesser models. We only need to compare the answers from GPT-4 to a lesser model to know that this is true.

The basic recipes for dealing with the bias-variance problem are as follows:

- Get as much data as possible.
- If we have a high bias problem (underfit), then use a bigger network, train longer, or change the architecture.
- The high variance problem is usually a more common and difficult one to solve. For high variance error (overfitting with the validation dataset), then consider using *regularization* in our network, get more data, or change the architecture. We will use regularization in many of our examples later on in the book and

will discuss them later, but if you come across terms like ridge regression, lasso, dropout, batch norm, or teacher forcing, then do not be overly concerned with knowing the exact details, they are just techniques to reduce high variance error and are relatively easy to implement in Keras.

Fortunately, Keras provides a tuner library to help pick the optimal set of parameters for our model. The Keras Tuner class has four tuners available: *Random Search, Hyperband, BayesianOptimization, and Sklearn.* The Sklearn tuner is used for Sklearn only; if we only use Keras machine learning models, then select one of the other three tuner classes: *Random Search, Hyperband, BayesianOptimization.*

The tuner class can tune many parameters at the same time, including the selection for the number of nodes and layers. In addition, model parameters, such as the learning rate, can also be tuned. We build our model and tell the tuner to tune the defined parameters. It will return the "best" model(s) to use along with a summary of the results.

The tuner serves an efficient way to select the hyperparameters. The main disadvantage in using a tuner is the long runtime, but it is quicker than trying to do it manually! Out of the three methods, Hyperband is the most efficient in resource allocation, and in practice, we should use either the Hyperband or the Bayesian algorithm.

We can see how the tuner works by using it on our Modified National Institute of Standards and Technology (UNIST) project. For convenience, the original code is repeated here followed by the new code tuned by Keras.

```
from keras import activations
import numpy as np
import keras
import tensorflow as tf
import matplotlib.pyplot as plt
from keras.models import Sequential
from keras.layers import Dense
from keras import activations
from keras.optimizers import Adam
from google.colab import files

# load mnist dataset
# 60,000 images for xTrain, 10,000 for xTest
mnist = tf.keras.datasets.mnist
(xTrain, yTrain), (xTest, yTest) = mnist.load_data()
#convert the 28x28 input matrix to a vector of length 784
xTrainFlattened = xTrain.reshape(len(xTrain),784)
xTestFlattened = xTest.reshape(len(xTest),784)

#set yTrain and yTest to 1 where the digit is 8 and
#0 for others
yTrain8 = np.where(yTrain!=8,0,1)
yTest8 = np.where(yTest!=8,0,1)

print("Count of yTrain = 8",(yTrain8 == 1).sum(),
" out of ",len(yTrain))
```

```
print("Count of yTest = 8",np.sum(yTest8 == 1),
" out of ",len(yTest))

#create the neural network as before
model = Sequential()
model.add(Dense(784, activation='relu',
        input_shape = (784,)))
model.add(Dense(784,activation='relu'))
 # the sigmoid activation will show
#the probability of each digit
model.add(Dense(1,activation='sigmoid'))
model.compile(optimizer=Adam(),
       loss='binary_crossentropy',
metrics=['accuracy']) # show the accuracy of prediction
model.fit(xTrainFlattened, yTrain8, epochs=5)
yPredict = model.predict(xTestFlattened)

# test the accuracy using the xTest
model.evaluate(xTestFlattened,yTest8)
```

We will tune the code above using the Keras tuner. Before we can use it, we may need to install the tuner package using the following command on Colab or our local terminal if we run the code locally on our PC:

```
!pip install keras-tuner
```

Once the library is installed, we can define a hypermodel by two methods. Either

- Creating a model builder function and passing a hyperparameter instance as a parameter to that function. The model builder function returns a compiled model and uses hyperparameters we define inline to tune the model. Once we have defined the search space, we can select a tuner class, such as `Hyperband` or `BayesianOptimization`, to start the search.
- Subclassing the HyperModel class of the Keras Tuner API.

In addition to these two methods, we can also use the two predefined HyperModel classes, HyperXception and HyperResNet, but since these are mainly used for computer vision applications only, we will not be addressing them here.

As an example of how to create a model builder function, we will use the tuner to choose the number of units, the activation function, and the learning rate in our UNIST application.

```
from keras import activations
import numpy as np
import keras
import tensorflow as tf
import matplotlib.pyplot as plt
from keras.models import Sequential
from keras.layers import Dense
from keras.layers import Dropout
from keras import activations
```

```
from keras.optimizers import Adam
from google.colab import files

#import the tuner
import keras_tuner

# load mnist dataset
# 60,000 images for xTrain, 10,000 for xTest
mnist = tf.keras.datasets.mnist
(xTrain, yTrain), (xTest, yTest) = mnist.load_data()
xTrainFlattened = xTrain.reshape(len(xTrain),784)
xTestFlattened = xTest.reshape(len(xTest),784)

#set yTrain and yTest to 1 where the digit is 8 and
#0 for others
yTrain8 = np.where(yTrain!=8,0,1)
yTest8 = np.where(yTest!=8,0,1)

print("Count of yTrain = 8",(yTrain8 == 1).sum(),
" out of ",len(yTrain))
print("Count of yTest = 8",np.sum(yTest8 == 1),
" out of ",len(yTest))

#create the neural network as before
def buildModel(hp):
model = Sequential()
model.add(Dense(784, activation='relu',
        input_shape = (784,)))

#define the search space
for i in range(hp.Int("num_layers", 1, 5)):
model.add(
Dense(
# Tune number of units separately.
units=hp.Int(f"layerUniti", min_value=2^4,
        max_value=2^10, step=2^3),
activation=hp.Choice("activation", ["relu", "tanh"]),))
if hp.Boolean("dropout"): model.add(Dropout(rate=0.25))
# choice of learning rates
learningRate = hp.Choice("lr", values=[1e-2, 1e-3, 1e-4])
model.add(Dense(1,activation='sigmoid'))
model.compile(optimizer=Adam(learning_rate=learningRate),
loss='binary_crossentropy', metrics=['accuracy'])
return model

buildModel(keras_tuner.HyperParameters())
tuner = keras_tuner.Hyperband(buildModel,
objective='val_accuracy',
max_epochs=10,
factor=3)
tuner.search_space_summary()
tuner.search(xTrainFlattened, yTrain8, epochs=5,
 validation_data=(xTestFlattened, yTest8))
tuner.results_summary()
```

```
Best val_accuracy So Far: 0.9947999715805054
Total elapsed time: 00h 30m 38s
Results summary
Results in ./untitled_project
Showing 10 best trials
Trial summary
Hyperparameters:
num_layers: 1
layerUnit0: 8
activation: relu
dropout: False
lr: 0.001
layerUnit1: 7
layerUnit2: 6
layerUnit3: 7
layerUnit4: 7
tuner/epochs: 10
tuner/initial_epoch: 4
tuner/bracket: 2
tuner/round: 2
tuner/trial_id: 0012
Score: 0.9947999715805054
Trial summary
Hyperparameters:
num_layers: 1
layerUnit0: 7
activation: relu
dropout: False
lr: 0.0001
layerUnit1: 6
layerUnit2: 8
layerUnit3: 6
layerUnit4: 8
tuner/epochs: 10
tuner/initial_epoch: 4
tuner/bracket: 2
tuner/round: 2
tuner/trial_id: 0015
Score: 0.992900013923645
Trial summary
Hyperparameters:
num_layers: 5
layerUnit0: 8
activation: relu
dropout: True
lr: 0.0001
layerUnit1: 6
layerUnit2: 6
layerUnit3: 8
layerUnit4: 6
tuner/epochs: 10
tuner/initial_epoch: 4
tuner/bracket: 1
tuner/round: 1
```

The optimizer took about 30 mins to complete all the epochs and completed its search for the "best" hyperparameters. I have included the first few best trials in the output above. We can either recreate our model using these parameters manually or retrieve it from the optimizer and save it down.

5.4 Customizations

In this section, we delve into the various ways Keras enables the customization of neural network architectures, focusing on the distinct approaches and the flexibility each offers for building tailored deep learning models.

In Keras, there are essentially three ways to use APIs to define a neural network: sequential, functional, and model subclassing. The sequential API is the easiest to implement, but it is the most restrictive with each layer only connecting to the subsequent layer in a linear fashion.

The functional API provides greater flexibility, allowing branches of layers, multiple inputs and outputs. A model such as ResNet or Inception is easily implemented using this method. In most cases, using the functional API is all we need to implement very sophisticated networks.

Lastly, model subclassing provides complete control over the training procedure: from customizing transfer function, custom layers, and models to monitoring and stopping the training process early using callbacks. Such flexibility comes with increasing complexity in coding and the removal of some useful utilities, such as print_summary(), etc.

5.5 Functional API

The main idea of the functional API is a way to build graphs of layers. Instead of the sequential model where each layer follows the previous one in a linear fashion, the functional API can handle models with nonlinear topology, shared layers, and multiple inputs or outputs. In the functional API, models are created by specifying their inputs and outputs in a graph of layers. That means that a single graph of layers can be used to generate multiple models. This flexibility is essential when we wish to implement the more advanced models, such as RNN, GAN, or CNN, in later chapters.

The key idea of the functional API is the use of one layer as a parameter input to another layer, so we can easily use one layer as input to several other layers and allow for branching in a network. To have several input layers into one layer, Keras provides a `concatenate` layer which merges the input layers into a single list to feed into the next layer.

Once the model is defined using the functional API, training, evaluation, and inference work exactly in the same way for models built using the functional API as for sequential models.

We will explain the use of the functional API by providing examples of how to create customized layers, models, and loss functions in the next few sections.

5.6 Custom Models

Custom models are created simply by feeding a layer with a preceding layer as an input. For example, let us define two inputs with two output models with a dense layer in between using the functional API. The code would be as follows:

```
import tensorflow as tf
import keras
from keras.layers import *
from keras.models import Sequential, Model
from keras.optimizers import Adam, RMSprop
import numpy as np
from keras.utils import plot_model

input1 = Input(shape=(32,),name="input1")
input2 = Input(shape=(128,),name="input2")
x1 = Dense(2, name = "dense1")(input1)
x = Dense(1,name = "dense2")(input2)
x2 = Dense(1,name = "dense2_2")(x)
concatted = Concatenate(name="concatted")([x1, x2])
model = Model(inputs=[input1, input2], outputs=concatted)

plot_model(model, show_shapes=True,
show_layer_names=True, to_file='modelconcat.png')
```

The `plot_model` function gives the following diagram which is a mapping of what we have just coded.

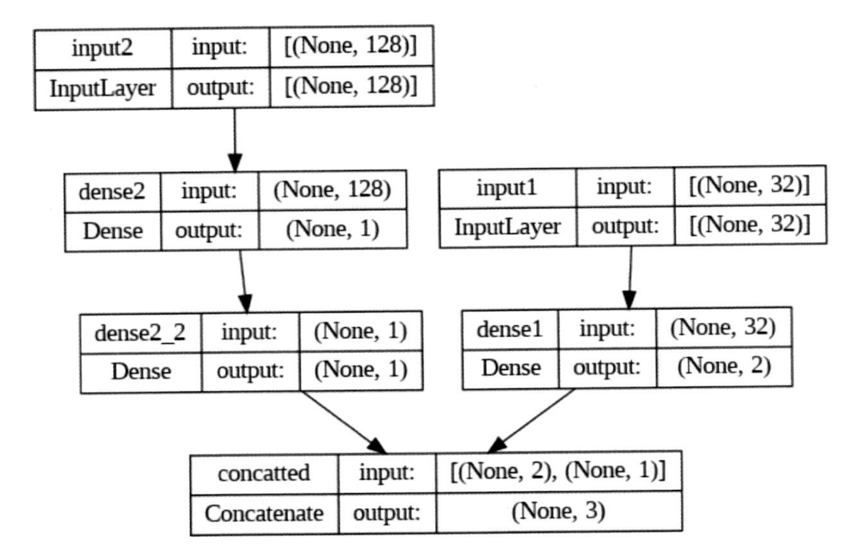

The functional API offers remarkable flexibility due to its ability to use a layer as an input to another layer. This feature allows us to create complex branching in the hidden layers and provides the flexibility to process various types of input data, such as numeric, categorical, and images, through different paths, merging them at the final layer. While we have omitted the activation function on the dense layer for clarity, it functions identically to a sequential network.

In later chapters, we will extensively use the functional API to introduce different network topologies. Regardless of how intricate the layers may seem, they are constructed in the same manner as described here.

Moreover, the functional API empowers us to override Keras's model base class, enabling the creation of customized models. A customized model groups layers with a set of input nodes, hidden layers, and output. This approach proves beneficial when the same group of layers needs to be repeated numerous times in a complex network. By defining a customized model, we can reuse the code without duplicating the layer structure.

For instance, let's consider the construction of a Siamese network using the functional API. A Siamese network comprises two identical subnetworks feeding into a comparator layer at the output. These subnetworks share the same architecture and parameters and are essentially mirror images of each other. If the weights of one subnetwork are updated, the weights of the other subnetwork are updated accordingly.

The Siamese network topology is often utilized to compute the degree of similarity between two items, such as images, words, or sounds. In this example, we will create a network to compare two images. This involves constructing two identical CNN subnetworks (referred to as Siamese sisters). Each network outputs a vector, which is then fed into a final comparative layer to assess the degree of dissimilarity.

```
import tensorflow as tf
from tensorflow.keras.layers import Input, Conv2D,
    MaxPooling2D, Flatten, Dense, Lambda
from tensorflow.keras.models import Model
import tensorflow.keras.backend as K

def initialize_base_network():
    #Define the base network (shared layers)

    input = Input(shape=(input_shape,),
      name="base_input")
    x = Conv2D(64, (3, 3), activation='relu')(input)
    x = MaxPooling2D((2, 2))(x)
    x = Conv2D(128, (3, 3), activation='relu')(x)
    x = MaxPooling2D((2, 2))(x)
    x = Flatten()(x)
    x = Dense(128, activation='relu')(x)
    return Model(inputs=input, outputs=x)

def euclidean_distance(vectors):
    #Define the Euclidean distance function
    x, y = vectors
    sum_square = K.sum(K.square(x - y), axis=1,
      keepdims=True)
    return K.sqrt(K.maximum(sum_square, K.epsilon()))

input_shape = (105, 105, 1)  # Example input shape

# Create the base network
base_network = initialize_base_network()

# Create the left input and point to
#the base network
input_a = Input(shape=input_shape, name="left_input")
processed_a = base_network(input_a)

# Create the right input and point to
#the same base network
input_b = Input(shape=input_shape, name="right_input")
processed_b = base_network(input_b)

# Calculate the distance between the
# two encoded inputs
distance = Lambda(euclidean_distance,
   output_shape=(1,),
   name="distance")([processed_a, processed_b])

# Create the Siamese network model
model = Model(inputs=[input_a, input_b],
  outputs=distance)

# Define the optimizer and compile the model
model.compile(loss="contrastive_loss",
   optimizer="adam")
```

```
# Now the model is ready to be trained with
# pairs of inputs and a label indicating
# their similarity
```

5.7 Model Selection

Model selection seems daunting as there are so many alternatives to choose from, but in practice, it is often the case that real-world requirements and constraints would help to guide us to the most appropriate model. As surprising as it may seem, it is often fine-tuning and experimentation that often take a long time to master and get right.

The considerations for model selection in this section are mainly based on learning and research-based requirements. Commercial projects often require complex and larger models, which are beyond the scope of this book. As an anecdote sidenote on model selection in a commercial environment, several software companies allow the user to run the data against hundreds of different models and parameters and choose the model with the best results. While this is a valid approach, it would not work well with less time and resources.

For many tasks, especially in computer vision and natural language processing, pretrained models are available. These can be a good starting point and can be fine-tuned on your specific dataset. At the very least, examining these pretrained models would give us a good idea of what a simplified architecture should be for our problem. For instance, if the pretrained model uses predominantly CNN or ResNet layers with good result, then a similar type of layers should be used for our project.

Speed requirement is an important factor to consider. For applications requiring real-time responses, like mobile or web applications, lightweight models are preferable, but for tasks where accuracy is paramount and response time is less critical, like medical image analysis, more complex models can be used to get better results.

The complexity of the model is a double-edged sword. More complex models have greater power and flexibility but are prone to overfitting, especially when the data size is limited. Techniques such as dropout, regularization, and data augmentation are employed to combat overfitting. Conversely, too simple a model might not capture the complexity of the data, resulting in underfitting and poor overall performance.

At the core of model selection is the specificity of the task. For instance, deep convolutional neural networks (CNNs) are often the go-to choice for image or speech recognition tasks, leveraging their ability to capture spatial hierarchies. On the other hand, sequence modeling tasks, like language translation or time-series prediction, benefit from the temporal dynamics captured by recurrent neural networks (RNNs), long short-term memory networks (LSTMs), or gated recurrent units (GRUs). In regression tasks where the goal is to predict numerical values, simpler DNN architectures might suffice, though complex data might still require the sophistication of CNNs or RNNs. Additionally, for generative tasks like image generation or style transfer, models like generative adversarial networks (GANs) or variational autoencoders (VAEs) are typically used.

The characteristics of the data also play a pivotal role in model selection. Large datasets can often support deeper and more complex networks, but this comes with increased computational demands. High-dimensional data, such as high-resolution images, may necessitate more sophisticated models to effectively capture underlying patterns. The quality and diversity of the training data are equally important; noisy, imbalanced, or unrepresentative datasets can lead to poor model performance, making data preprocessing and augmentation crucial.

In summary, while the selection of an appropriate model can be a complex task, understanding the specific requirements and constraints of our project is crucial in guiding you toward the most effective and efficient model choice.

5.8 Model Depth and Complexity

Each layer in a DNN extracts a level of abstraction of the data. In general, more layers allow the network to learn more complex patterns. For instance, in a CNN that is used for image recognition, initial layers might detect edges, while deeper layers could identify more complex structures, like shapes or specific objects.

The complexity and volume of your data are key indicators when choosing the model depth and complexity. Simple patterns and smaller datasets often require fewer layers, as a deep network might overfit, learning noise rather than useful patterns. Conversely, complex data such as high-resolution images or intricate language structures might benefit from deeper architectures.

Although we have not discussed the transformer layer in this book due to its complexity, using transformer layers in a neural network architecture is particularly advantageous for certain types of problems, especially those involving sequential data or requiring the understanding of long-range dependencies. The transformer model, introduced in the seminal paper "Attention Is All You Need" by Vaswani et al., has revolutionized the field of natural language processing (NLP) and is increasingly being adapted for other applications, such as computer vision.

In conclusion, the depth and complexity of a neural network model should be carefully tailored to the specific characteristics of our data and the nature of the task at hand, balancing the ability to learn intricate patterns with the risk of overfitting, to achieve optimal performance.

5.9 Neural Networks Applications

We will now present some applications which were state-of-the-art models a few years ago. The main purpose of presenting my own implementation of these research projects is to illustrate two important points:

- The reader should, hopefully, be able to follow and understand the code now.
- Gain insights into the subtle but crucial considerations researchers must account for when designing innovative models.

These applications, demand significant computing resources. All of them take days, if not several weeks, to run on a normal PC with a decent GPU, so be prepared to invest the time to experiment.

In most cases, it is a trade-off between time and the quality of training. For example, suppose we want to train the agent to play the Space Invaders game "well," we would need to define, monitor, and understand how well the model is learning and when to adjust or stop training. Here are some strategies to effectively monitor the performance:

- Track Cumulative Rewards per Episode: The most direct measure of performance in reinforcement learning is the cumulative reward obtained in each episode. Plotting this over episodes gives a clear view of whether the agent is improving.
- Evaluate Loss During Training: Keep an eye on the loss of the neural network during the training process. A decreasing loss trend indicates that the model is learning, but if the loss plateaus or increases, it might signal issues with the training process.
- Use a Running Average: Due to the inherent variability in reinforcement learning, it's useful to look at a running average of the rewards over a set number of episodes (e.g., the last 100 episodes). This smooths out the noise and gives a clearer trend.
- Test in a Fixed Environment: Periodically test our agent in a controlled environment where the starting conditions are fixed. This can give us a more consistent basis for comparison.
- Performance Thresholds: Set performance benchmarks or thresholds. For example, if the agent achieves a certain score or survives a certain number of frames consistently, it can be a sign of adequate learning.
- Visual Inspection: Occasionally watch the agent play the game. This can give our qualitative insights into its learning progress and strategy development.
- Log and Review Training Data: Regularly log important metrics, like rewards, loss, epsilon values, and specific actions taken. Reviewing these logs can help identify patterns or issues.
- Compare Against Baselines: If available, compare our agent's performance against known baselines or benchmarks for the game. This gives a context to the performance.
- Use Validation Episodes: Separate training and validation episodes can be useful. Train the agent for a number of episodes, then test it on different episodes without learning (i.e., no weights update during these test episodes). This helps in understanding how well the model generalizes.
- Early Stopping: Implement an early stopping mechanism. If the performance does not improve or starts to degrade over a certain number of episodes, stop the training to prevent overfitting or wasting resources.

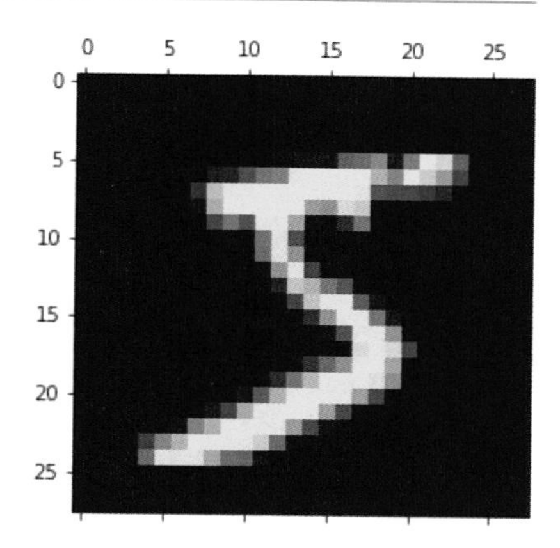

Figure 5-4 MNIST 28 × 28 matrix of digit 5

5.10 Dense Network: Detection of Handwritten Digits Using MNIST Dataset

Just to illustrate this point further, let us consider the problem of recognizing handwritten digits. This problem at first seems nothing like the problem in the previous example, but with some minor code changes to digitize the images, we can use the same type of neural network to solve it.

We are going to use the popular MNIST dataset. The MNIST dataset is so widely used in machine learning that it has been included as part of Keras's standard built-in datasets. The dataset consists of 60,000 images of digits ranging from zero to nine, where each digit is represented as a 28 × 28 matrix of grayscale pixels. In addition to the 60,000 training images, the dataset includes 10,000 images for testing purposes.

An example of an MNIST digit is shown in Figure 5-4. Since this is a grayscale image, each pixel represents a brightness value from 0 (black) to 255 (white). The yTrain and yText arrays contain the actual number of the digit itself, so for the image, yTrain=5. To use the model, we need to convert the 28 × 28 array of each input image to a vector of length 28*28=784 using the method reshape on the NumPy array so that the neural network can accept it. The particular type of neural network we are using for the example requires this input format, so we need to do the conversion. The input layer differs for different networks, so there is no one fixed format, but they all required numerical inputs.

The code for the project is shown below:

```
import numpy as np
import keras
import tensorflow as tf
import matplotlib.pyplot as plt
from keras.models import Sequential
from keras.layers import Dense
from keras.optimizers import Adam
from google.colab import files
```

```
# load mnist dataset
# 60,000 images for xTrain, 10,000 for xTest
mnist = tf.keras.datasets.mnist
(xTrain, yTrain), (xTest, yTest) = mnist.load_data()
xTrainFlattened = xTrain.reshape(len(xTrain),784)
xTestFlattened = xTest.reshape(len(xTest),784)

#create the neural network as before
model = Sequential()
model.add(Dense(784, activation='relu',
       input_shape = (784,)))
model.add(Dense(784, activation='relu'))
model.add(Dense(10,activation='softmax'))
model.compile(optimizer=Adam(),
loss='sparse_categorical_crossentropy',
metrics=['accuracy'])

model.fit(xTrainFlattened, yTrain, epochs=5)
# test the accuracy using the xTest
model.evaluate(xTestFlattened,yTest)

313/313 [==============================] -
1s 4ms/step - loss: 0.1969 - accuracy: 0.9597
[0.19687649607658386, 0.9596999883651733]
```

As we can see, the model setup is very similar to the previous example, although two differences are worth noting that the model uses "sparse_categorical_crossentropy" as the loss function and "softmax" for the activation of the output layer.

Do not worry about these parameters or what they mean for the time being. We will discuss these topics in the next few sections. The point to note here is how the same type of neural network could be used for two seemingly unrelated applications. Using just a few lines of code, we have managed to identify handwritten digits with 96% accuracy.

If we (correctly) feel that the whole machine learning process is relatively simple up to now, it is only because the complexity of machine learning has been wrapped up in the Keras library. This is an important point; in practice, if we are only interested in using machine learning in applications, then we only need to master a few important things:

- Understand the whole process and the tools from beginning to end.
- Learn to preprocess the input data to feed the models effectively.
- Select and configure different models.
- Know how to evaluate the model to improve its accuracy.

However, while Keras simplifies the machine learning process, it is important not to underestimate the complexity of the built-in libraries and to recognize the complex algorithms and computations that operate behind the scenes, making such simplicity possible.

5.11 RNN Network: Modeling an AutoRegressive Integrated Moving Average (ARIMA) Process

So far, our discussions have revolved around the simplest type of deep neural network known as the feedforward neural network (FNN). In this network, information flows only in the forward direction: from the input layer, through various hidden layers, and to the output layer. The backpropagation algorithm is used to update the weights of the feedforward network, and it does not involve creating loops or cycles in the physical connections between nodes.

Feedforward networks are primarily employed for supervised learning tasks, such as object classification or pattern recognition. They excel at creating classification boundaries between different classes of objects, particularly when the data to be learned is not time dependent nor sequential. However, for time series analysis or models that require self-learning, more complex network structures are necessary. In such cases, loops in the network are used to enforce dependencies between the current state and previous states of the network.

Recurrent networks are typically used for time series analysis, such as an encoder-decoder network for sequence-to-sequence prediction problems. Additionally, other networks, like generative and diffusion models, are utilized to generate additional instances similar to the training data. Many renowned applications, such as ChatGPT, DALL-E, and other large language models (LLMs), currently employ variants of generative models with massive training data and computational power, resulting in impressive effects.

One of the simplest alternative neural network structures is a recurrent neural network (RNN). RNN is widely used to model time series and sequential tasks. A sequential task involves input and output data sequences, such as text streams, stock prices, or video clips, which can be modeled using RNNs.

If we look at Figure 5-5, the small diagram on the left shows the recurrent nature of the network. That is, it is a set of layers, with each input fed by the output of the preceding hidden layer. The diagram on the right displays the exact structure of the network. $\{x_1, x_2, x_3 \ldots, x_t\}$ represents the sequence of input values which feed into

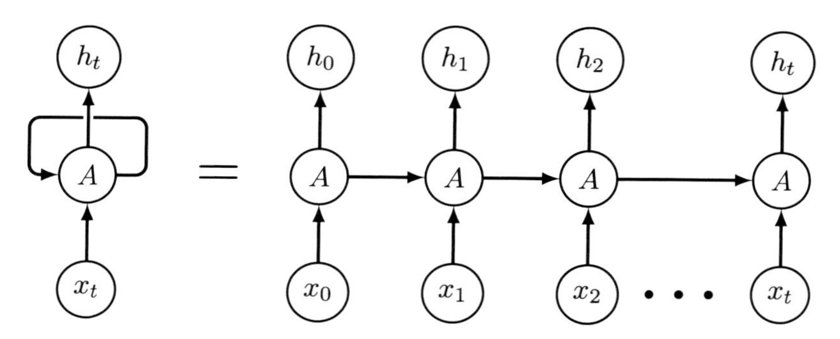

Figure 5-5 Diagram showing a naive RNN network

the hidden neural network layer denoted by A. The node A represents the memory state at time t such that

$$A(t) = f(w_1 X_t + w_2 A(t-1))$$

In simpler terms, the current memory state at time t, denoted as $A(t)$, is a nonlinear function f of the input value at time t, denoted as X_t, and the previous state value at time $t-1$. Each $h_0, h_1, \ldots, h_t$ represents the output value at each step of the sequence. Notably, h_t corresponds to the final output of the sequence. Structurally, node A can contain one or multiple layers, but for ease of implementation, all nodes A typically share the same activation function and hyperparameters.

So far, we have not defined the specific structure for node A. However, it's worth mentioning that RNN networks often encounter the vanishing gradient problem in practice. To mitigate this issue, a type of RNN called long short-term memory (LSTM) is commonly used.

The structure of an RNN (recurrent neural network) enables it to effectively model both linear and nonlinear relationships. One significant advantage of using an RNN model is its capability to handle different time series without the need to select alternative (linear and nonlinear) models. Moreover, an RNN model possesses memory, making it particularly suitable for modeling autoregressive (AR) models, where the value of the output Y_t at time t depends on previous q values.

To illustrate the effectiveness of RNNs in time series analysis, we will compare its forecasting performance against the ARIMA(p, d, q) model using 20 years of the Standard & Poor's 500 closing prices.

To begin with, let us model the time series using an ARIMA(p, d, q) model. This model has three parameters:

$$Y_t = \sum_{i=1}^{p} \omega_{t-i} Y_{t-i} + \sum_{j=1}^{q} \theta_{t-j} \epsilon_{t-j}$$

In this equation, Y_t is the predicted value at time t. The model combines three key components:

- Autoregressive (AR) terms represented by $\sum_{i=1}^{p} \omega_{t-i} Y_{t-i}$, where p is the number of lag observations included in the model; ω_{t-i} are the coefficients for the lags of the series at times $t-1, t-2, \ldots, t-p$; and Y_{t-i} are the lagged values of the series.
- Differencing order d, which is the number of times the data have had past values subtracted (not explicitly shown in the equation).
- Moving average (MA) terms represented by $\sum_{j=1}^{q} \theta_{t-j} \epsilon_{t-j}$, where q is the size of the moving average window, θ_{t-j} are the coefficients for the lagged forecast errors in the prediction equation, and ϵ_{t-j} are the lagged forecast errors at times $t-1, t-2, \ldots, t-q$.

The ARIMA model thus captures the dynamics of a time series through a combination of these autoregressive and moving average terms, adjusted by the level of differencing to ensure stationarity.

We can use statistical tests, such as the Augmented Dickey-Fuller (ADF) and Canova-Hansen (CH), or (partial) autocorrelation tests (PACE and ACE) to choose the parameters' values.

Stock prices are usually nonstationary, so differencing using the d parameter is often necessary to stationary. In practice, this condition is hard to achieve, and finding the right values for p, d, and q often requires experimentation.

We will use 90% of the dataset for training and 20% for forecasting.

Let's implement the ARIMA model using Python's statsmodels library:

```
import pandas as pd
import numpy as np
import matplotlib.pyplot as plt
from statsmodels.tsa.arima.model import ARIMA
from sklearn.metrics import mean_squared_error
from statsmodels.graphics.tsaplots import plot_acf, plot_pacf

data = pd.read_csv('sp500_data.csv',
        index_col='Date',parse_dates=True)
ts_data = data['Close']
train_size = int(len(ts_data) * 0.8)
train, test = ts_data[:train_size], ts_data[train_size:]

# Plot ACF and PACF
plot_acf(ts_data, lags=20)
plot_pacf(ts_data, lags=20)
plt.show()

P = 1
d = 1
q = 1

model = ARIMA(train, order=(p,d,q))
model_fit = model.fit()
```

Now, let's model the same time series using an RNN for comparison. Instead of using ARIMA, we can use RNN to do the same task:

```
import numpy as np
import pandas as pd
from sklearn.preprocessing import MinMaxScaler
from tensorflow.keras.models import Sequential
from tensorflow.keras.layers import Dense, LSTM
import matplotlib.pyplot as plt

data = pd.read_csv('sp500_data.csv')
prices = data['Close'].values.reshape(-1, 1)

# Scale the data to values between 0 and 1
scaler = MinMaxScaler(feature_range=(0, 1))
scaled_prices = scaler.fit_transform(prices)
```

```
# Define the number of time steps for the RNN
time_steps = 30

# Create sequences for the RNN
X, y = [], []
for i in range(len(scaled_prices) - time_steps):
X.append(scaled_prices[i : i + time_steps])
y.append(scaled_prices[i + time_steps])

X, y = np.array(X), np.array(y)

# Split the data into training and testing sets
train_size = int(0.8 * len(X))
X_train, X_test = X[:train_size], X[train_size:]
y_train, y_test = y[:train_size], y[train_size:]

model = Sequential()
model.add(LSTM(50, return_sequences=True,
            input_shape=(X_train.shape[1], 1)))
model.add(LSTM(50))
model.add(Dense(1))

model.compile(optimizer='adam',
      loss='mean_squared_error')
model.fit(X_train, y_train, epochs=50, batch_size=32,
       verbose=1)

# Predict using the test data
predicted_prices = model.predict(X_test)
predicted_prices = scaler.
       inverse_transform(predicted_prices)

# Transform the original test data back
    # to the original scale
y_test = scaler.inverse_transform(y_test)

# Calculate Mean Squared Error (MSE)
mse = np.mean((predicted_prices - y_test) ** 2)
print("Mean Squared Error:", mse)

# Plot the predictions against the actual values
plt.figure(figsize=(12, 6))
plt.plot(y_test, label="Actual Prices", color="blue")
plt.plot(predicted_prices, label="Predicted Prices",
     color="red")
plt.xlabel("Time")
plt.ylabel("S&P 500 Index Price")
plt.title("S&P 500 Index Prediction using RNN")
plt.legend()
plt.show()
```

5.12 LSTM Network: BachBot

BachBot is a research application written by Feynman Liang [4] at Cambridge University, UK, to compose classical musical chorales in the style of composer J.S. Bach using an LSTM network. The code presented here is the basic version of the model inline with Feynman's published paper and his thesis. The original implementation was written in PyTorch and Lua and is a more sophisticated implementation with more functionalities, including harmonization. The implementation in this book is not a straight conversion of the code but written from scratch using Keras. I am grateful to Feynman for useful discussions and for allowing me to use some of his MusicXML utility code.

5.12.1 Background

Johann Sebastian Bach's music is often described as mathematical due to its structure, precision, and the way it adheres to specific musical forms and rules. Bach, a master of counterpoint, skillfully interwove multiple independent melodies that harmonized with each other. This technique, especially evident in his fugues, requires meticulous planning and a deep understanding of harmonic and melodic structures. The mathematical aspect lies in how these independent melodies fit together in a precise, logical way, often following strict rules.

J.S. Bach's 400+ chorales are hymn tunes written to be sung by a congregation in a German Protestant church service. In Bach's time, these hymns were often harmonized for four voices (soprano, alto, tenor, and bass) for performance by a choir or used as the basis for more complex compositions. The text of chorales is usually in German and reflects Lutheran religious themes, often based on biblical scriptures or liturgical texts.

Bach's chorales are known for their rich harmonic texture and complex counterpoint. Each voice part is melodically interesting and contributes to the overall harmonic structure. An example of his chorales is shown below. Note that the four staffs are notes for the four voices: soprano (top), alto, tenor, and bass (bottom). Additionally, each chorale contains a series of phrases which Bach delimited using fermatas. Choosing an LSTM (long short-term memory) network for a project like BachBot, which is focused on generating music in the style of Johann Sebastian Bach, makes sense due to several key characteristics of LSTM networks that are particularly well suited for handling music and sequential data in general. Specifically

- Handling of Sequential Data: Music is inherently sequential, with each note or chord depending on what came before. LSTM networks excel at processing and generating sequential data because they can maintain information in "memory" over long sequences, which is crucial for capturing the structure and progression in music.

I. Ach bleib' bei uns, Herr Jesu Christ

Figure 5-6 J.S. Bach—BWV 649—Ach bleib bei uns, Herr Jesu Christ. Notice phrasing using fematas [4]

- Long-Term Dependencies: Bach's compositions often contain long-term dependencies, where themes or motifs are developed and revisited over time. LSTMs are specifically designed to capture such long-term dependencies in sequences, unlike traditional neural networks which struggle with long-range data dependencies.
- Learning Patterns in Time: LSTMs can learn patterns in time-series data. Bach's music, with its complex rhythms, harmonies, and counterpoints, presents a rich temporal structure that LSTMs can learn and model.
- Variability in Sequence Length: Musical pieces can vary significantly in length, and LSTMs can handle this variability effectively. They can be trained on sequences of varying lengths and generate sequences that are not fixed in size.
- Flexibility in Modeling Different Elements: LSTMs can be used to model various aspects of music, such as melody, harmony, rhythm, and dynamics, by learning from a diverse range of musical features encoded in the input data.

5.12.2 Preprocessing

Once the network topology is chosen, Feynman encodes Bach's music for digitization in a very particular format. Firstly, all chorales are transposed to the key C-major for major scores and A-minor for minor ones. The C-major or A-minor

Figure 5-7 Sample of a J.S. Bach chorale [4]

key is chosen to make the encoding process easier by removing all the sharps and flats in the original key signature.

Time quantization is also performed. Each note duration is transformed into a multiple of a semibreve duration, neglecting changes in timing (e.g., ritardandos), dynamics (e.g., crescendos), and additional notation (e.g., accents, staccatos, legatos). The pitch of each note is represented in MIDI specification from 0 to 127 representing ten octaves. Although articulation marks are removed, the fermata, denoted by the symbol (.), is retained to help the model learn phrasing.

Finally, akin to bar lines in standard music notation, a frame delimiter

(|||)

is used to segment the music into timing measures.

The end result of the preprocessing stage is a series of tokens shown in Figure 5-8. A (true) boolean value in the token denotes a tied note.

The generation of a corpus of token files for the 400+ Bach chorales is where we start. Although this process only needs to be performed once, it completes quickly. Therefore, the code is kept in the program to run each time, avoiding additional logic that could clutter the main program.

Figure 5-8 An example of the tokenized file for the J.S. Bach score BWV 133.6

```
START
(59, True)
(56, True)
(52, True)
(47, True)
|||
(59, True)
(56, True)
(52, True)
(47, True)
|||
(.)
(57, False)
(52, False)
(48, False)
(45, False)
|||
(.)
(57, True)
(52, True)
(48, True)
(45, True)
|||
END
```

5.12.3 Model Implementation and Training

The Keras model for BachBot is shown below. Feynman's thesis detailed the hyperparameters for PyTorch. Fortunately, the corresponding Keras parameters are compatible, which makes the implementation much easier. One particular aspect of the training process that is nonstandard is the use of teacher forcing for the RNN network.

5.12.4 Teacher Forcing

Teacher forcing is a training methodology where, instead of using the model's own predictions as inputs during training, the model is provided with the actual or expected output from the previous time step. In other words, during training, the network is "forced" to consider the correct output (as provided by the teacher, which in this case is the training data) at each step, instead of its own predictions.

In standard RNN training without teacher forcing, the predicted output from the previous time step is fed back into the network as input for the next time step. Instead of using the RNN's own predictions from the previous step, teacher forcing uses the actual previous output from the training dataset.

The advantages of teacher forcing include faster convergence and stability during training. However, a major drawback of teacher forcing is the discrepancy between training and inference. During training, the model always sees the correct previous output, but at inference, it must generate sequences based on its own predictions. This difference can lead to issues such as exposure bias, where the model may not perform well during inference.

```
def teacher_forcing_encode(corpus, files, time_steps):
# Tokenize the corpus
corpus_tokens = corpus.split('@')
tokenizer = Tokenizer(filters='', split='@',
  lower=False)
tokenizer.fit_on_texts(corpus_tokens)

# Convert tokens to numeric values
sequences = tokenizer.texts_to_sequences(corpus_tokens)
d_tokens = np.array([item for sublist in sequences
                for item in sublist])

# Unique tokens and vocabulary size
unique_tokens = tokenizer.word_index
vocabulary_size = len(unique_tokens)

# Prepare X and Y
num_files = len(files)
X = []
Y = []

# Find start and end indices for each file
start_indices = [i for i, x in enumerate(corpus_tokens)
                if x == 'START']
end_indices = [i for i, x in enumerate(corpus_tokens)
                if x == 'END']
end_token = 0
for i in range(num_files):
file_seq = d_tokens[start_indices[i]:end_indices[i]]
X.append(np.array(file_seq[:-1]))
Y.append(np.array(file_seq[1:]))

x_chunks_list = [split_and_pad_sequence(seq,
  time_steps,end_token) for seq in X]
X = [chunk for sublist in x_chunks_list for chunk
    in sublist]
X = np.array(X)
y_chunks_list = [split_and_pad_sequence(seq,
  time_steps, end_token) for seq in Y]
Y = [chunk for sublist in y_chunks_list for chunk
     in sublist]
Y = np.array(Y)

# Convert labels to one-hot encoding
```

```
    Y_train_one_hot = to_categorical(Y,
      num_classes=vocab_size+1)

    return X, Y_train_one_hot, tokenizer, corpus_tokens
```

The code above does set up the conditions for implementing teacher forcing during the training of the LSTM model, although the actual implementation of teacher forcing happens within the model's training process itself.

The function teacher_forcing_encode prepares the training data (X_train and Y_train) in a format suitable for teacher forcing. In this setup, for each sequence in X_train, the corresponding sequence in Y_train is the same sequence but shifted by one time step, providing the "next token" as the target for each input sequence.

The model is trained using X_train and Y_train. During training, the LSTM model receives a sequence from X_train and learns to predict the next token in the sequence. The correct next token is always the corresponding element in Y_train. This means the model is trained with the actual next token from the training data as the target for each time step, which is the essence of teacher forcing.

The create_bachbot_lstm_model function constructs the LSTM model using the specified parameters. When this model is trained with the X_train and Y_train data, it implicitly uses teacher forcing because of how the training data is structured.

I found that the implementation works well when producing music when the initial feed from Bach's choral works but not so when asked to produce long passages of music. Scheduled sampling is a recent alternative training method for resolving this discrepancy, but it is a matter of experimentation and is beyond the scope of the project.

5.12.5 BachBot Model

The model used by BachBot is a fairly standard LSTM model to generate music sequences.

```
def create_bachbot_lstm_model(num_layers, rnn_size,
    embedding_dim, seq_length, dropout_rate,
    vocab_size):
    model = Sequential()
    model.add(Embedding(input_dim=vocab_size,
    output_dim=embedding_dim, input_length=seq_length))
    model.add(BatchNormalization())
    model.add(Dropout(dropout_rate))
    for _ in range(num_layers):
        model.add(LSTM(rnn_size, return_sequences=True))
        model.add(BatchNormalization())
        model.add(Dropout(dropout_rate))
        model.add(TimeDistributed(Dense(vocab_size+1,
            activation='softmax')))
```

using these following constants as parameters as specified in the research paper: $num\_layers = 3$, $rnn\_size = 256$, $embedding\_dim = 32$, $seq\_length = 128$, and $dropout\_rate = 0.3$.

Essentially, this is a sequential RNN model with 128 time steps. The rnn_size, sets to 256, defines the number of units in each LSTM layer, which is three in this case. Before the input data enters the LSTM network, it is embedded to a lower dimension of 32. Although embedding is used to reflect the configuration of the paper, in practice it does not affect the result significantly and can be removed to improve the speed if needed.

When we set return_sequences=True for an LSTM layer in a neural network, it means that the layer will return the full sequence of outputs for each time step in the input sequence. This is in contrast to the default setting (return_sequences=False), where the LSTM layer only returns the output of the last time step.

With return_sequences=True, the LSTM layer produces an output for each time step in the input data, preserving the temporal sequence information. This is essential when subsequent layers in the model also expect time series data, such as in sequence-to-sequence models, or when we are stacking multiple LSTM layers in this case.

The TimeDistributed layer is used in combination with a dense layer. This setup is used so that after processing the sequence data through LSTM layers, we want to make a decision for each note at each time step of the sequence, rather than collapsing the entire sequence into a single output at the end.

The dense layer with a softmax activation is applied to each time step output of the preceding LSTM layer. This setup allows the model to make predictions of the next musical note at each step in the sequence. After training, the result can be exported to an xml file format and read and played using a software package, such as MuseScore. An example of the BachBot's composition using the program is shown in Figure 5-9.

Figure 5-9 BachBot's musical composition

Generative Models 6

So far in this book, we have introduced several architectures predominantly to classify objects. These models aim to minimize the differences between the predicted object and the trained ones to make accurate predictions. Recently, there have been huge interests in machine learning, not to classify objects but to generate new contents based on learned ideas. This class of machine is referred to as generative AI.

6.1 Variational Autoencoders

Generative AI refers to a class of artificial intelligence systems that are capable of generating new content that resembles some input data they were trained on. These systems use machine learning algorithms, particularly deep learning, to learn patterns and structures from existing data and then use that knowledge to create new content.

As of 2023, the primary architectures in generative AI include generative adversarial networks (GANs), variational autoencoders (VAEs), and diffusive models. The three architectures are very different, and each model is preferred for alternative applications. The diffusive model class is the newest and is the most complicated, but it requires a very large dataset and is extremely computing intensive and time-consuming to train.

The main idea behind VAEs is to sample data from a learned distribution of trained data. A VAE model consists of two parts: an encoder and a decoder. The job of the encoder is to encode data into a smaller set of data, often referred to as the latent space, using a feedforward neural network as before. However, the encoder output is not an encoded vector but the parameters of a distribution. Often, the output parameters are the mean μ and the standard deviation σ that represent a compressed Gaussian distribution of the input data. The latent space can be thought of as encoding the most important features of the data. To fine-tune our newly

P. Hua, *Neural Networks with TensorFlow and Keras*,
https://doi.org/10.1007/979-8-8688-1020-6_6

generated data, we often need these features to be independent from one another. This sometimes can be done using disentanglement techniques, which are beyond the scope of this book.

The decoder part of the network then draws a sample from the latent space and passes through another neural network to get an approximation of a reconstructed version of the data. This allows us to sample from the approximated distribution to generate new data.

To aid understanding of a VAE network, think of an example of Identikit. A likeness of a person's face constructed from descriptions given to police uses a set of transparencies of various facial features that can be combined to build up a picture of the person sought. The encoder basically generates the set of facial features for the decoder to choose from.

More formally, in probabilistic terms, the encoder outputs the parameters set λ for the Gaussian probability density function $q_\lambda(z|x)$, where z is the latent variable for input x. The decoder takes the latent input z and output the trained PDF $p(\hat{x}|z)$ from which new data is generated.

The loss function that we want to optimize is called ELBO (evidence lower bound) rather than the common mean squared error (MSE) which does not make sense for the purposes of VAEs. This loss function guides the VAE to learn a balance between accurately reconstructing the input data and maintaining a structured latent space.

The ELBO function has two parts:

- A Log-Likelihood: This is used to minimize the difference between the input data and the trained output.
- A KL Divergence: This metric measures the difference between the latent distribution and the reconstructed distribution. It takes in two distributions as arguments and outputs the KL divergence.

To see how a VAE works in practice, we can use it to create new faces by training the model using the CelebFaces dataset images, which can be downloaded from the Internet.

6.1.1 Preprocessing

The images are normalized to [0,1] as input tensors using the following line of code:

```
validationData = utils.image_dataset_from_directory(
    os.path.dirname("C:/AI/Data/VAE/validation"),
    image_size=(img_height,img_width),batch_size=batch_size)
dsValidation = validationData.map(lambda x, y: x/255.0)
```

6.1.2 VAE Architecture

The encoder, a convolutional neural network, encodes input images into a latent space representation. It outputs two vectors: z_mean and z_log_var. The sampling layer samples from the latent space using the reparameterization trick, which is then passed to the decoder which decodes the latent space representation back into images.

```
def compute_loss(self, x):
    z_mean, z_log_var = self.encoder(x)
    z = self.sampling((z_mean, z_log_var))
    x_reconstructed = self.decoder(z)
    reconstruction_loss = tf.reduce_mean(
    tf.keras.losses.binary_crossentropy(x, x_reconstructed)
    ) * 64 * 64 * 3
    # fig, axes = plt.subplots(1, 2, figsize=(10, 5))

    # Display the first image in the first subplot
    # axes[0].imshow(tf.keras.backend.eval(x[1].numpy()))
    # axes[1].imshow(tf.keras.backend.eval(x_reconstructed[1]
                .numpy()))
    # plt.tight_layout()

    # tf.print("reconstruct:",reconstruction_loss)
    kl_loss = -0.5 * tf.reduce_sum(1 + z_log_var -
        tf.square(z_mean) - tf.exp(z_log_var), axis=1)
    kl_loss = tf.reduce_mean(kl_loss)
    #tf.print("kl loss:",kl_loss)
    vae_loss = reconstruction_loss + kl_loss
    return vae_loss
```

In the code above, the compute_loss function in our script is designed to calculate the loss for a variational autoencoder (VAE). This function implements the key components of the VAE loss, which includes both the reconstruction loss and the Kullback-Leibler (KL) divergence:

- z_mean, z_log_var = self.encoder(x): This line encodes the input data x using the VAE's encoder to get the mean, z_mean, and log variance, z_log_var, of the latent variables.
  z = self.sampling(z_mean, z_log_var)
  The sampling layer uses z_mean and z_log_var to generate a sample z from the latent space.
  x_reconstructed = self.decoder(z)
  The sampled latent variables z are then decoded back into the reconstructed data. The reconstruction loss is computed using binary cross-entropy between the original input x and the reconstructed output x_reconstructed.
  This loss measures how well the VAE can reconstruct the input data from the latent variables. It is scaled by the dimensions of the input data (64 * 64 * 3), to account for the total number of pixels and color channels in the input images.

Figure 6-1 Left to right: Generated faces after 50 and 1000 epochs

- KL Divergence Loss
  kl_loss = –0.5 * tf.reduce_sum(1 + z_log_var – tf.square(z_mean) – tf.exp(z_log_var), axis=1)
  This line computes the KL divergence between the approximate posterior defined by z_mean and z_log_var and the prior distribution which is assumed to be a standard normal distribution. The KL divergence acts as a regularizer, encouraging the distribution of the latent variables to be close to a standard normal distribution.
  This is crucial for ensuring a well-structured and meaningful latent space.
  kl_loss = tf.reduce_mean(kl_loss)
  The KL divergence is averaged over the batch.
- Total VAE Loss
  vae_loss = reconstruction_loss + kl_loss
  The total loss for the VAE is the sum of the reconstruction loss and the KL divergence. This combined loss function is what the VAE will try to minimize during training.

The output of generated faces from the VAE networks is shown in Figure 6-1. More realistic faces will be produced with longer training and more experimentation with hyperparameters. Figure 6-2 shows the original and reconstructed face images.

One major disadvantage of VAEs is that they can have a problem known as posterior collapsing where the resultant images are blurred. There are techniques to minimize this, but generally a GAN will produce higher-quality images at the cost of training time.

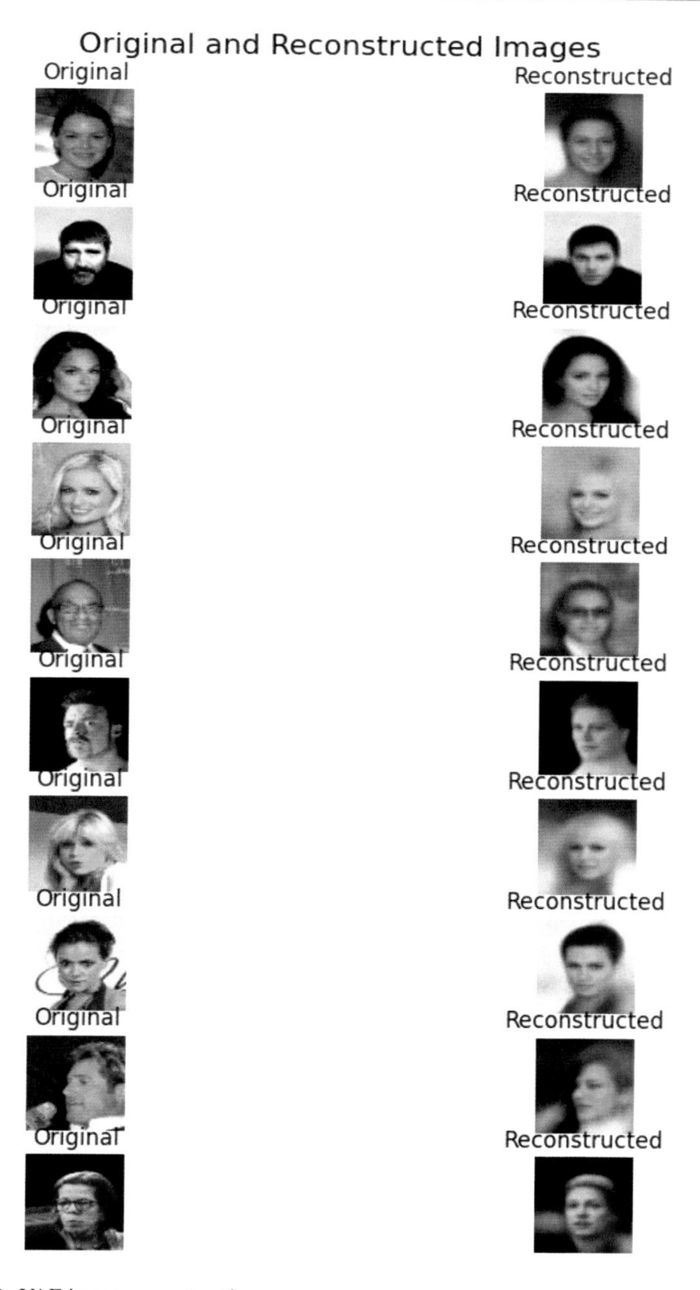

Figure 6-2 VAE image reconstruction

The Python code supplied provides one interesting tool for visualizing the distribution of data points in the latent space of a neural network that is worth noting:

```
# Function to plot the latent space
def plot_latent_space(model, data, num_samples=1000):
    sampled_data = data.take(num_samples)
    images = []
    for img, _ in sampled_data:
    images.append(img)
    images = tf.concat(images, axis=0)
    z_mean, _ = model.encoder.predict(images)
    tsne = TSNE(n_iter=300, perplexity=30,
       learning_rate=200)
    z_mean_reduced = tsne.fit_transform(z_mean)
    plt.scatter(z_mean_reduced[:,0],z_mean_reduced[:,1])
    plt.xlabel('Dimension 1')
    plt.ylabel('Dimension 2')
    plt.title('Latent Space Visualization')
    plt.show()
```

The purpose of this code is to provide a visual representation of how data points are distributed in the latent space of the VAE after encoding. It can help you gain insights into the structure and separability of the latent space, which is useful for assessing the quality of the learned representations and understanding how well the VAE has disentangled the underlying factors of variation in the data.

The provided code defines a function to plot the latent space of a variational autoencoder (VAE) or similar neural network model. Here's a breakdown of what the code does.

The function takes three arguments:

Model This argument represents the VAE model or a similar model that has an encoder capable of encoding data into a latent space.

Data This argument represents a TensorFlow dataset (data) containing input data samples.

Num Samples This argument specifies the number of samples to use for visualization. The default value is set to 1000.

The function starts by sampling a subset of data points from the dataset. It takes the first *numsamples* samples from the dataset.

For the sampled data, it encodes the data using the VAE's encoder (model.encoder) to obtain the mean (*zmean*) of the latent space representation. Then, it applies t-Distributed Stochastic Neighbor Embedding (t-SNE), a dimensionality reduction technique, to reduce the dimensionality of the latent space representation from its original dimensionality to a two-dimensional space.

After reducing the dimensionality, it creates a scatter plot of the two-dimensional latent space. Each point in the scatter plot represents a data point, and its position is determined by the reduced latent space representation obtained using t-SNE.

Finally, it displays the scatter plot, allowing you to visualize the distribution and clustering of data points in the latent space.

The purpose of this code is to provide a visual representation of how data points are distributed in the latent space of the VAE after encoding. It can help you gain insights into the structure and separability of the latent space, which is useful for assessing the quality of the learned representations and understanding how well the VAE has disentangled the underlying factors of variation in the data.

6.1.3 Morphing Images

Morphing images in the latent space from one image to another is a fascinating application using VAE. Morphing in a variational autoencoder (VAE) involves smoothly transitioning between two input images by navigating the latent space of the VAE. Normally, morphing between two images involves linear interpolation between the two vectors representing the two from-and-to images, but the interpolation can follow other paths depending on the desired transitional effect. Using a pretrained VAE network for morphing is straightforward using the following steps:

- Load our pretrained VAE model.
- Encode the source and target images to obtain their latent representations.
- Linearly interpolate between the latent vectors and decode them to produce a sequence of morphed images.
- Finally, we display or save the morphed images.

The resulting images will smoothly transition from the source to the target, creating a morphing effect. We can adjust the number of frames and interpolation method to control the smoothness and speed of the morphing. Using our pretrained VAE code, we can perform morphing from one face to another. The result is shown in Figure 6-3

```
import numpy as np
import matplotlib.pyplot as plt
from keras.models import load_model

# Load your pre-trained VAE model
vae = load_model('celeb_vae_model.h5')

# Encode source and target images
source_image = ...  # Load your source image
target_image = ...  # Load your target image

latent_source, _ = vae.encoder.predict(source_image)
latent_target, _ = vae.encoder.predict(target_image)

# Specify the number of frames for morphing
num_frames = 10
```

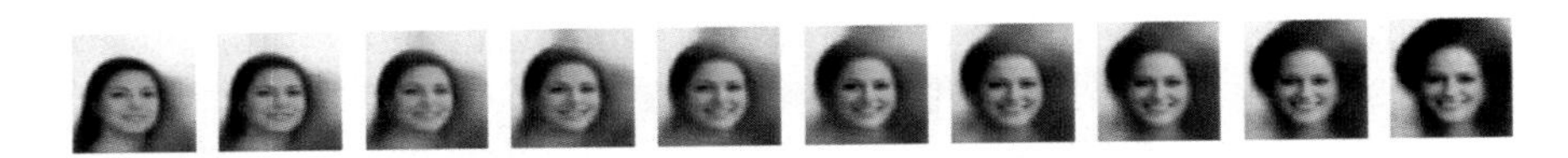

Figure 6-3 Transition from one image to another in latent space

```
# Linear interpolation in latent space
morphed_images = []
for alpha in np.linspace(0, 1, num_frames):
interpolated_latent = (1 - alpha) *
                latent_source + alpha * latent_target
decoded_image = vae.decoder.predict(interpolated_latent)
morphed_images.append(decoded_image)

# Display or save morphed images
for i, image in enumerate(morphed_images):
plt.imshow(image.squeeze(), cmap='gray')
plt.axis('off')
plt.title(f'Interpolation Step {i}')
plt.show()
```

6.1.4 Feature Disentanglement

As we perform morphing using the above code, it is clear that the latent space using VAE is highly entangled. Features are added to the intermediate steps in an uncontrolled fashion. Disentangled VAEs have become a prominent area of research in machine learning due to their potential for creating more interpretable and controllable generative models.

Having disentangled latent space is potentially very useful. The latent space learned by these models can have practical applications in areas such as image manipulation, style transfer, and data generation. By changing the values of individual dimensions in the latent space, we can control specific attributes of the generated data. For example, in image generation, we might have separate dimensions for factors like pose, color, and style.

Currently, achieving perfect disentanglement is still an area of research. However, there are various approaches to produce reasonable disentanglement in VAEs, including

- Conditional VAE (CVAE): A CVAE extends the VAE to condition the latent space on certain variables or features. You can design the model to generate a latent space representation that explicitly encodes these features. This is often used in conditional image generation tasks.
- Disentangled VAE (β-VAE, Factor-VAE): These are variations of VAEs that aim to encourage the latent dimensions to capture independent and interpretable

factors of variations in the data. While they may not directly encode separate features, they can help create more interpretable and disentangled representations.
- InfoGAN: Information Maximizing Generative Adversarial Networks (InfoGANs) are a type of VAE-GAN hybrid that introduces an auxiliary network to explicitly maximize the mutual information between a subset of the latent variables and the generated data. This encourages those variables to capture specific attributes of the data.
- Joint-VAE: Joint-VAE combines a VAE with a clustering algorithm to encourage the model to learn a more disentangled representation.

To see how factor disentanglement works in practice, we can modify our VAE model to implement the Factor-VAE, a model proposed by Hyunjik Kim and Andriy Mnih in their paper "Disentangling by Factorising." Implementing Factor-VAE using Keras involves modifying a standard variational autoencoder (VAE) loss function slightly to include a factor disentanglement term in the loss function. The disentanglement term encourages the VAE to learn a more structured and disentangled latent representation as below. The remaining code for the VAE program remains as before:

```
def compute_loss(self, x):
    z_mean, z_log_var = self.encoder(x)
    z = self.sampling((z_mean, z_log_var))
    x_reconstructed = self.decoder(z)
    reconstruction_loss = tf.reduce_mean(
    tf.keras.losses.
    binary_crossentropy(x, x_reconstructed)
    ) * img_width * img_height * 3

    kl_loss = -0.5 * tf.reduce_sum(1 + z_log_var -
            tf.square(z_mean) - tf.exp(z_log_var), axis=1)
    kl_loss = tf.reduce_mean(kl_loss)

    # Factor disentanglement term
    factor_disentanglement_loss = tf.
      reduce_mean(tf.abs(z_mean))

    vae_loss = reconstruction_loss + kl_loss +
                    factor_disentanglement_loss
    return vae_loss
```

Once we have encoded the data, it will not be obvious to determine what each encoded dimension actually represents. In general, Factor-VAE, and other VAE models, cannot produce an independent axis of feature representation, so if we examine a particular dimension, it is unlikely that the single dimension is a unique feature. The metric for measuring disentanglement is not straightforward but is needed for commercial work. For educational purposes however, it is sufficient to view visually the effect along feature changes to guess the resultant purpose of each latent dimension. The code below does this by looping through all dimensions and modifying one dimension at a time to see the result:

```
import tensorflow as tf
import numpy as np
import matplotlib.pyplot as plt

# Load your trained FactorVAE model and encoder here
factorvae = tf.keras.models.load_model('vae_model.h5')
encoder = tf.keras.models.load_model('encoder_model.h5')

# Load or generate an input data point for reconstruction
input_data = np.random.randn(1, input_dim)

# Encode the input data
latent_representation = encoder.predict(input_data)

# Get the dimensionality of the latent space
latent_dim = latent_representation.shape[1]

# Initialize an empty array to store
# reconstructed data along each dimension
reconstructed_data_per_dim = []

# Loop through each dimension and
# modify it while keeping others fixed
for dim_index in range(latent_dim):
# Create a copy of the original latent representation
modified_latent_representation =
    latent_representation.copy()

# Set a new value for the current dimension
# You can modify this value based
# on your desired transformation
new_value = 2.0  # Example: Set a new value

# Modify the current dimension
modified_latent_representation[0, dim_index] = new_value

# Decode the modified latent representation
# to generate reconstructed data
reconstructed_data = factorvae.decoder.
           predict(modified_latent_representation)

# Append the reconstructed data to the list
reconstructed_data_per_dim.append(reconstructed_data)

# Plot the original data and reconstructed
# data along each dimension
plt.figure(figsize=(4 * latent_dim, 4))
for dim_index in range(latent_dim):
plt.subplot(1, latent_dim, dim_index + 1)
plt.imshow(reconstructed_data_per_dim[dim_index][0].
           #reshape(image_shape), cmap='gray')
```

```
    plt.title(f'Dimension {dim_index}')
    plt.axis('off')

    plt.tight_layout()
    plt.show()
```

6.2 CartoonGAN

The project "CartoonGAN: Generative Adversarial Networks for Photo Cartoonization" was proposed by Yang Chen et al. [5] as a solution for transforming photos of real-world scenes into cartoon-style images. There have been many previous attempts to do the same, but they tend to use pair images and cartoons which are very time-consuming to generate. Often, this process would involve an artist drawing the required cartoon to pair up the original image.

These following points are taken from their published paper, which states the main contributions of the model:

- (1) We propose a dedicated GAN-based approach that effectively learns the mapping from real-world photos to cartoon images using unpaired image sets for training. Our method is able to generate high-quality stylized cartoons, which are substantially better than state-of-the-art methods. When cartoon images from individual artists are used for training, our method is able to reproduce their styles.
- (2) We propose two simple yet effective loss functions in GAN-based architecture. In the generative network, to cope with substantial style variation between photos and cartoons, we introduce a semantic loss defined as an L1 sparse regularization in the high-level feature maps of the VGG network. In the discriminator network, we propose an edge-promoting adversarial loss for preserving clear edges.
- (3) We further introduce an initialization phase to improve the convergence of the network to the target manifold. Our method is much more efficient to train than existing methods.

6.2.1 GAN

GANs consist of two neural networks, a generator and a discriminator, which are trained simultaneously through adversarial processes. The generator generates data that is as realistic as possible, while the discriminator evaluates this data, trying to distinguish between real and generated (fake) data.

The goal of GANs is to generate data, often images but also other data types, that are indistinguishable from real data. As training progresses, the generator improves in producing more realistic data, while the discriminator becomes better at telling real from fake.

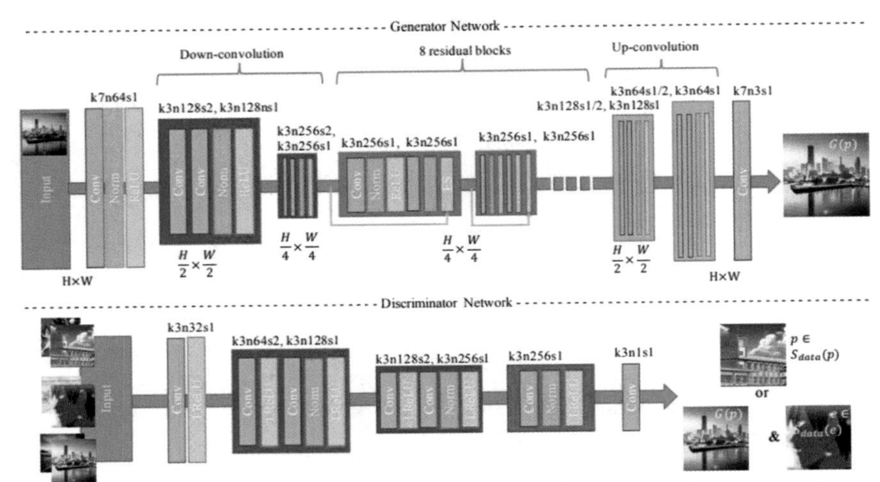

Figure 2. Architecture of the generator and discriminator networks in the proposed CartoonGAN, in which k is the kernel size, n is the number of feature maps and s is the stride in each convolutional layer, 'norm' indicates a normalization layer and 'ES' indicates elementwise sum.

Figure 6-4 The CartoonGAN architecture [5]

GANs are primarily used in tasks involving data generation, such as image and video generation, style transfer, data augmentation, and more.

The architecture for CartoonGAN in Figure 6-4 is relatively standard. Both generator and discriminator networks are CNN networks as shown below, which we have implemented in Keras. In their original paper, the authors use pictures taken from the film *Spirited Away*, which for copyright reasons, we cannot include in our dataset. However, we will now discuss the tools needed to replicate the data used in the research paper from Flickr and the movie for the interested reader.

6.2.2 Data Preparation

The research paper used images from two sources: Flickr and Spirited Away. Both of these sources have limited distribution rights, so we would need to create our own set of data using the procedure below to train the model.

For Flickr, Jeff Heaton [6] provided an excellent and easy-to-use utility to automate the download which have been included in the source file. However, we need to edit the config_flickr.ini file to search for the images required. In this instance, I have downloaded images related to the theme "nature" of size 256 × 256 into the local directory /home/philip/AI/Data/CartoonGAN/PhotoFlickr.

```
[FLICKR]
id = your account id
secret = your secret id
[Download]
path = /home/philip/AI/Data/CartoonGAN/PhotoFlickr
```

```
search = Nature
prefix = Nature
update_minutes = 1
license = 0,1,2,3,4,5,6,7,8,9,10
max_download = 100000
sources_file = sources.csv
[Process]
process = True
crop_square = True
min_width = 256
min_height = 256
scale_width = 256
scale_height = 256
image_format = jpg
```

Running the flickr-download.py code should download approximately 4000–5000 images into the local directories.

Capturing frames from the movie Spirited Away is trickier. We would need to have the movie in .mov format on the local drive, then download FFmpeg, and use the following command to extract frames every four seconds (or whatever interval we need):

```
ffmpeg -i input.mov -r 0.25 output_%04d.jpg
```

We would need to remove the blank images at the beginning and end. These should be placed into a separate directory.

The next step is to crop these images to 256 × 256 using the function crop_images_in_directory() in pre_process.py. These cropped images are then augmented using augment_images(), which flips the images left to right to create more images.

CartoonGAN requires smooth images, which can be done by running smooth.py with the appropriate directory for the captured cartoon images. Ensure that the output directory is different to the unsmoothed images.

We should now have three separate directories containing the photos, cartoons, and smoothed cartoons; each image of size 256 × 256 is nearly ready to be fed into the model. The directories for these images should be set in main.py under three separate directories. For example:

```
photo_paths_norm = '~/AI/Data/CartoonGAN/PhotoNorm/'
cartoon_paths_norm = '~/AI/Data/CartoonGAN/CartoonNorm/'
smoothed_paths_norm = '~/AI/Data/CartoonGAN/CartoonSmoothNorm/'
```

6.2.3 Preprocessing CartoonGAN

The images are normalized in create_dataset() to a range of [−1,1]. This range is needed because we are using $tanh$ as the activation function instead of $softmax$. tanh is generally used in hidden layers or in recurrent network structures to normalize and regulate the flow of data, whereas softmax is used in the output layer for classification tasks to represent the likelihood of the input belonging to each class. However, the difference in result is marginal in this case, and either activation function could be used.

```
def create_dataset(image_directory, batch_size=1):
    image_paths = [os.path.join(image_directory, fname)
    for fname in sorted(os.listdir(image_directory))
    if os.path.isfile(os.path.join(image_directory, fname))
        and fname.lower().endswith('.jpg')]

    # Define a function to process the images
    dataset = tf.data.Dataset.
       from_tensor_slices(image_paths)
    dataset = dataset.map(lambda x:
                load_and_preprocess_image(x, 256, 256))
    dataset = dataset.batch(batch_size)
                .prefetch(tf.data.AUTOTUNE)
return dataset

def load_and_preprocess_image(path,target_height,
  target_width):
    image = tf.io.read_file(path)
    image = tf.image.decode_jpeg(image, channels=3)
    image = tf.image.convert_image_dtype(image,
                dtype=tf.float32)
    image = (image - 0.5) * 2  # Scale to [-1, 1]
    return image
```

The AUTOTUNE parameter is used to optimize data transfer in batches in TensorFlow. When loading and preprocessing large datasets, the efficiency of operations like data fetching, batching, and preprocessing can significantly impact training speed. AUTOTUNE allows TensorFlow to automatically determine the optimal number of batches to process in parallel and the best configuration for other performance-related settings.

The tensors created by the code above are suitable for the generator and discriminator models defined in the project. However, CartoonGAN uses a VGG network for training content in the generator, and this model needs tensors of size 224×224 in the range of [0,1], so our tensors need to be converted. This is done in the function preprocess_for_vgg().

```
def preprocess_for_vgg(image):
    # Resize to VGG input size
    image = tf.image.resize(image, (224, 224))
    image = (image + 1) / 2  # convert [-1,1] to [0,1]
    return preprocess_input(image)  # Normalize for VGG
```

6.2.4 The Discriminator Model

Of the two, the discriminator model is the simpler one. The topology of the neural network follows closely the discriminator structure stated in the research paper [5].

The loss function needs more explaining, although, in general, it follows the form of a typical GAN model except for the loss due to the smoothing of images.

```
def loss(real_images, g_generated_images, smoothed_images):
    real_output = discriminator(real_images,training=True)
    fake_output = discriminator(g_generated_images,
                  training=True)
    edge_output = discriminator(smoothed_images,
                  training=True)
    real_losstf.keras.losses.BinaryCrossentropy
        (from_logits=False)
        (tf.ones_like(real_output), real_output)
    fake_loss= tf.keras.losses.BinaryCrossentropy
        (from_logits=False)
        (tf.zeros_like(fake_output), fake_output)
    edge_loss = tf.keras.losses.BinaryCrossentropy
        (from_logits=False)
        (tf.zeros_like(edge_output), edge_output)
    discriminator_loss = real_loss +fake_loss +edge_loss
    return discriminator_loss
```

The function takes three parameters: real_images are images from the Flickr dataset, g_generated_images are the fake images generated by the generator, and smoothed_images are the cartoon images which have been smoothed. At first glance, it is counterintuitive to consider that we do not need to feed the same images into the discriminator. However, it makes sense when we realize that the model is targeting style and feature transfer, and the loss due to content, fake_loss, is only one of the three losses being considered.

Setting training=True indicates that the model should run in training mode, which might include behaviors like dropout.

The loss calculation calculates three different types of losses for the images generated by the discriminator using binary cross-entropy, a common loss function for binary classification tasks:

- real_loss: Measures how well the discriminator can identify real images. It compares the discriminator's output for real images (real_output) with a tensor of ones (tf.ones_like(real_output)), representing the correct classification of real images.
- fake_loss: Measures how well the discriminator can identify fake images generated by the generator. It compares the discriminator's output for fake images (fake_output) with a tensor of zeros (tf.zeros_like(fake_output)), representing the correct classification of fake images since all generator images are fake.
- edge_loss: This is an additional loss term for the smoothed images. This is one of the features of CartoonGAN where real cartoons are distinguished from the real images by sharp edges, so the edge_loss can be considered as a style loss.

6.2.5 The Generator Model

The generator role is to produce realistic cartoons to fool the discriminator. In a typical GAN, the initial images are usually just random noise. For CartoonGAN, the initialization stage relies on the VGG network to initialize the network weights to produce reasonably realistic images so that the second phase mainly deals with style transfer.

Hence, for the initialization phase, no discriminator is involved. We simply make use of the deep layer "block4_conv3" in VGG-19 to teach the generator to learn important features of the images. Once this is completed, the adversarial nature of the GAN starts. We see the logic for the two phases in the main loop.

```
for cartoon_images, photo_images, smoothed_images in
zip(cartoon_dataset,   photo_dataset, smoothed_dataset):
    if is_initialization:  # Generator Initialization Phase
        gen_loss, content_loss = generator.init_train(
         photo_images,content_lambda)
        disc_loss = 0
        adversarial_loss = 0
        style_loss = 0
    else:  # Adversarial Training Phase
        gen_loss, content_loss, adversarial_loss,
           style_loss =
            generator.train(discriminator.discriminator,
            photo_images, cartoon_images,
            g_adv_lambda, content_lambda,
            style_lambda)
        disc_loss = discriminator.train(generator.generator,
            photo_images, smoothed_images)

    with writer.as_default():
        # Log discriminator metrics
        tf.summary.scalar('Discriminator Loss',
         disc_loss, step=n)
        tf.summary.scalar('Generator Loss',
         gen_loss, step=n)
        tf.summary.scalar('Content Loss',
            content_loss, step=n)
        tf.summary.scalar('Adversarial Loss',
            adversarial_loss, step=n)
        tf.summary.scalar('Style Loss', style_loss, step=n)
        n += 1
        print(f"epoch, n, g_loss:gen_loss,
            d_loss:disc_loss,
            c_loss:content_loss,
            adv_loss:adversarial_loss,
            style_loss:style_loss")
    checkpoint_manager.save()
    test_picture = next(iter(photo_dataset.take(1)))
    print(test_picture.shape)
    preprocess.generate_and_save_images(
         generator.generator,
         test_picture, epoch)
```

In the initial phase, the model returns the generator loss, which is the content loss produced from the VGG-19 network.

In the second phase, both the generator and discriminator are trained. This time, however, several losses are computed: content, adversarial, and style losses are all calculated.

Content loss ensures the generated image content resembles the real images, adversarial loss makes the generated images indistinguishable from real ones, and style loss ensures the generated images have the desired artistic style.

Each loss is associated with a hyperparameter that adjusts its weight, thereby determining its importance in the overall loss calculation. In practice, by varying the ratios, we can produce a spectrum of images—from realistic looking photos to abstract blobs of color cartoons.

Through these two phases, the CartoonGAN model adeptly balances content fidelity and artistic stylization, enabling the creation of a diverse range of images, from photorealistic to highly stylized cartoons.

```
def train(discriminator, real_photos, cartoons,
        g_adv_lambda,content_lambda, style_lambda):
    with tf.GradientTape() as tape:
      generated_images = generator(real_photos,
          training=True)
      d_g_generated_images = discriminator(generated_images,
                                  training=False)
      cont_loss = content_lambda *
                    content_loss(real_photos,
                    generated_images)
      adv_loss = g_adv_lambda *
                    adversarial_loss(d_g_generated_images)
      s_loss = style_lambda *
                    style_loss(cartoons, generated_images)
      total_generator_loss = adv_loss +cont_loss +s_loss

    gradients = tape.gradient(total_generator_loss,
                    generator.trainable_variables)
    optimizer.apply_gradients(zip(gradients,
                    generator.trainable_variables))
    return total_generator_loss,cont_loss,adv_loss,s_loss
```

The style loss included in the code uses the Gram matrix to help with style transfer.

A Gram matrix is a mathematical representation used to measure the correlation between different feature maps in a convolutional neural network (CNN). It is calculated by multiplying a matrix of feature maps by its transpose. If we have a set of feature maps, which can be thought of as different filters applied to an image, the Gram matrix gives us a measure of how these feature maps activate together, essentially capturing the texture or style information of the image.

The output of the network after the initialization and the second phase are shown below in Figure 6-5. As can be seen, the image after the initialization phase is quite close to the original photo but contains some stripey artifacts from the CNN network.

Figure 6-5 Left to right: Original image, after initialization and final output

The second phase image is much more cartoon like. As stated below, the appearance can be changed by altering the parameters of the losses and rerunning the model as the quality or style of the output is achieved or, more likely, our patience runs out!

6.3 Stable Diffusion

In concluding our chapter on generative models, we turn our attention to the state-of-the-art Stable Diffusion AI model. Released in 2022, Stable Diffusion is an advanced deep learning model designed for text-to-image conversion, utilizing diffusion techniques. A notable distinction from its contemporaries like DALL-E and Midjourney, which operate as cloud services, is Stable Diffusion's ability to function on consumer PCs equipped with a GPU boasting at least 4 GB of RAM.

A key aspect of this model is its utilization of techniques discussed earlier in this chapter, specifically latent embedding, variational autoencoders (VAEs), and a convolutional neural network architecture known as U-Net. Originally developed for biomedical image segmentation, U-Net has been effectively adapted for generating images with fine-grained details, which are crucial for Stable Diffusion.

The training process of diffusion models is centered around the strategic addition and subsequent removal of Gaussian noise from training images. The Stable Diffusion model operates in three distinct phases:

- First, a variational autoencoder (VAE) compresses the image into a latent space, capturing the semantic relationships within the image.
- During the forward diffusion stage, Gaussian noise is incrementally added to this latent representation. The U-Net architecture then comes into play, de-noising the diffused representation in the latent space, aiming to produce an output that aligns with the intended text association.

- Finally, the VAE decoder reconstructs the image from the latent space back into pixel space.

The integration of a diffusion model alongside a VAE might raise questions. One could hypothesize about embedding the text prompt with the image directly in the latent space and generating the image using solely a VAE network. However, employing a diffusion network has several advantages.

Diffusion models have demonstrated superior capabilities in generating images of high quality and resolution. This excellence stems from their adeptness at modeling complex, high-dimensional data distributions, enabling them to capture and reproduce fine details and a wide variety of image features.

These models benefit from a more stable training process. By transforming the data into Gaussian noise in a controlled manner and subsequently learning to reverse this process, diffusion models exploit the well-understood properties of Gaussian noise. This approach provides significant control over the image generation process, which can be finely tuned by adjusting the levels of noise and the sampling methodology.

The VAE encoder and decoder are similar to the previous example; hence, we will focus on the two remaining components of the model: injecting and removing noise and the U-Net topology.

6.3.1 Text Embedding in Stable Diffusion

In the context of generative models, like Stable Diffusion, a latent space is a high-dimensional space where complex data (like images) are represented in a compressed, abstract form. This latent space captures the essential characteristics and variations of the data, making it easier for the model to manipulate and generate new data instances.

In Stable Diffusion, images are first encoded into this latent space using a variational autoencoder (VAE). The VAE compresses an image into a lower-dimensional latent representation, which retains the key features and semantic content of the original image. This process facilitates the manipulation of the image at a more abstract level, which is computationally efficient and conceptually powerful.

The latent space is designed to have desirable properties, like continuity and smoothness, meaning small changes in the latent representation result in small and predictable changes in the output image. This is crucial for the gradual transformation and generation processes in the model.

For text prompt integration, the model receives textual descriptions or prompts as input. These prompts describe the desired output image, such as "a sunny beach" or "a futuristic cityscape."

The challenge lies in mapping these text prompts to the latent space. This is typically achieved using a text encoder, which converts the text prompt into a feature

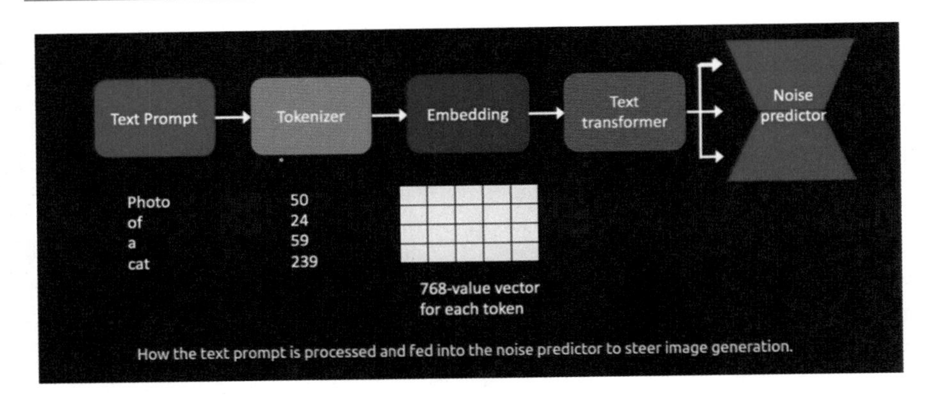

Figure 6-6 Stable diffusion architecture [7]

vector in the same latent space or a compatible one. This encoding captures the semantic meaning of the text.

The encoded text prompt and the latent image representation are combined or aligned in such a way that the text influences the image generation process. This could involve conditioning the diffusion process on the text features or directly modifying the latent image representation based on the text features.

The integrated text prompt guides the de-noising process of the diffusion model. As the model iteratively removes noise from the latent representation, it is influenced by the text features, leading it to generate an image that corresponds to the textual description.

The text conditioning process in Stable Diffusion is shown graphically in Figure 6-6. The output of the text transformer is used after word embedding. It is used to generate conditioning text that describes the desired attributes, features, or content that the diffusion model should produce. For example, if you want to generate images of "red apples," the text transformer might generate the prompt "a picture of a red apple."

6.3.2 Gaussian Noise Injection and Removal

Adding Gaussian noise to an image is straightforward in TensorFlow. If we have an image as a NumPy array, we can simply generate a random tensor from the normal distribution and add the two tensors together:

```
import tensorflow as tf
import numpy as np
import matplotlib.pyplot as plt

image = image.astype('float32') / 255.0

def add_noise(img, noise_level):
    noise = tf.random.normal(shape=img.shape,
```

```
                mean=0.0, stddev=noise_level)
    noisy_img = img + noise
    # Clip the values to maintain them within [0, 1]
    noisy_img = tf.clip_by_value(noisy_img,
          clip_value_min=0.0, clip_value_max=1.0)
    return noisy_img

noise_level = 0.1  # Adjust the noise level as needed
noisy_image = add_noise(image, noise_level)
```

The amount of noise at each stage and the total number of stages are important to the model and the nature of the dataset. More time steps generally allow for a more gradual and controlled diffusion process, potentially leading to better-quality image generation. However, this comes at the cost of increased computational complexity and longer training times. Typical values for the number of time steps range from 25 to several thousands. More stages and carefully tuned noise levels can lead to higher fidelity in generated images, but also require more complex and potentially slower models. Datasets with more complex and varied images might benefit from more stages and a carefully designed noise schedule.

Generally, we would start a low number of time steps, observe the quality of the result, and then increase gradually according to a noise schedule. The noise schedules can be linear or nonlinear. A linear schedule increases the noise linearly over time, whereas a nonlinear schedule might increase noise more quickly or slowly at different stages. Nonlinear schedules are often used because they can better model the data distribution of natural images.

The code shown above is actually a simplification of the noise injection process in a forward diffusion process. The actual code used is coded slightly differently to allow better control of noise injection. Denoting the current image at step t as x_t, then we create the image at $t + 1$ as a combination of the original image x_t and Gaussian noise, where the proportion of noise versus the original data is determined by the time step t and the parameter β_t

Mathematically, the image x_{t+1} at time $t + 1$ is expressed in a form $x_{t+1} = \sqrt{1 - \beta_t} x_t + \beta_t \epsilon$ where β_t is the noise level at time step t and ϵ is a noise term sampled from a normal distribution $N(0, 1)$. A nice property for Gaussian noise is that it is additive, meaning that instead of calculating x_t from x_{t-1}, x_t could be calculated from x_0 directly as follows:

Let $\alpha_t = 1 - \beta_t$, then

$$x_1 = \sqrt{\alpha_1}.x_0 + \epsilon\sqrt{1 - \alpha_1}$$

$$x_2 = \sqrt{\alpha_2}(\sqrt{\alpha_1}.x_0 + \sqrt{1 - \alpha_1}.\epsilon) + \sqrt{1 - \alpha_2}.\epsilon$$

$$x_2 = \sqrt{\alpha_2 \alpha_1}.x_0 + \epsilon(\sqrt{1 - \alpha_2} + \sqrt{\alpha_2(1 - \alpha_1)})$$

The additive variance noise property for a Gaussian noise means that we can add the two terms together to give

$$\epsilon(\sqrt{1-\alpha_2} + \sqrt{(\alpha_2(1-\alpha_1))}) = \epsilon\sqrt{1-\alpha_1\alpha_2}$$

such that we can express x_t in terms of x_0 succinctly as follows:

$$x_t = x_0 \prod_{i=1}^{t} \sqrt{\alpha}_i + \epsilon \sqrt{1 - \prod_{i=1}^{t} \alpha_i}$$

So in Python, we have two ways of implementing noise injection. The original equation is clearer to code, so I prefer to use it.

```
import numpy as np

def forward_diffusion(x_0, beta_schedule, t):
    x_t = x_0
    for i in range(1, t+1):
     beta_t = beta_schedule[i-1]
     noise = np.random.normal(size=x_0.shape)
     x_t = np.sqrt(1 - beta_t) *
              x_t + np.sqrt(beta_t) * noise
    return x_t
```

Using either method produces a sequence of noise-added images as shown in Figure 6-7.

This process of adding controlled noise is referred to as reparameterization trick. In the context of Stable Diffusion, it is thus a crucial technique for the effective training of the diffusion process. It allows the model to learn how to add noise in a controlled manner and, more importantly, how to reverse this process during the generation phase. This technique helps in maintaining differentiability of the model, which is essential for training deep learning models using gradient descent.

The process of de-noising a noisy image involves the reverse process of the forward diffusion. The de-noising process, which is the core of image generation

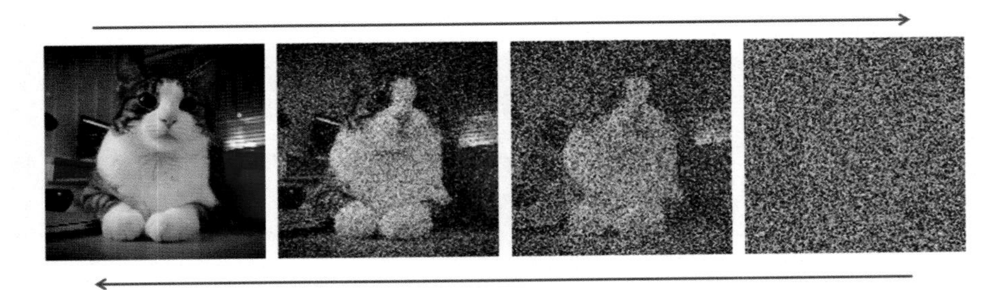

Figure 6-7 Adding noise to an image [8]

in Stable Diffusion, involves iteratively removing this noise to reconstruct an image or generate a new one.

Begin with a highly noisy image, which could be the result of the forward diffusion process applied to an original image, or it could be a randomly initialized noise image if we are generating new images. Similar to the forward diffusion process, the de-noising process is iterative and goes through the time steps in reverse order, starting from the last time step and going back to the first. At each time step, the model predicts the noise that was added at that particular step during the forward process and subtracts it from the current image state.

A neural network, typically a U-Net architecture, is used for the de-noising task. This network is trained to predict the noise that was added to the image at each time step. During the reverse process, the U-Net takes the noisy image and possibly additional conditioning information (like text embeddings in text-to-image models) as input and outputs an estimate of the noise that needs to be removed.

The estimated noise is then subtracted from the noisy image, resulting in a less noisy version of the image. This process is repeated at each time step, progressively reducing the noise and bringing the image closer to a clear state.

By the end of this reverse process, the image has undergone a series of refinements, and, ideally, we are left with a clear, coherent image that either closely resembles the original (in case of image reconstruction) or represents a new image generated based on the provided conditioning (like a text description).

The efficiency and quality of de-noising depend heavily on the training and architecture of the U-Net model, as well as on the accuracy of the noise prediction at each step. A high-level extract for the de-noising process is shown below. It assumes we have a pretrained model which can make predictions on the noise.

```
def denoise_image(noisy_image, model, num_timesteps,
                  conditioning_info=None):
    current_image = noisy_image
    for time step in reversed(range(num_timesteps)):
        # U-Net model predicts the noise
        predicted_noise = model.predict(current_image,
                          time step, conditioning_info)
        # Subtract the predicted noise from the image
        current_image = current_image - predicted_noise
    return current_image
```

At this stage, it is worth to pause the explanation on the de-noising process and discuss the architecture of the U-Net model and how its network is used to predict the noise process.

6.3.3 The U-Net Model

The U-Net model is a type of convolutional neural network (CNN) that is particularly effective for tasks like image segmentation and, as seen in recent advancements, for de-noising in diffusion models.

Image segmentation is a process in computer vision where an image is divided into multiple segments (sets of pixels, also known as superpixels). The goal of image segmentation is to simplify or change the representation of an image into something that is more meaningful and easier to analyze. It is used to locate objects and boundaries (lines, curves, etc.) in images. In contrast to image classification, which provides a label to the whole image, segmentation adds labels to each pixel or superpixels, thereby splitting (or segmenting) the image into significant parts.

There are different types of segmentation: semantic, instance, and panoptic, for example. They differ mainly in the way the pixels are classified. Semantic segmentation categorizes the pixels into predefined classes; instance segmentation categorizes each class into different instances of the same class. Instance segmentation in the context of Stable Diffusion models would involve generating or modifying an image based on text input, where the model not only recognizes and manipulates different objects within the image but also distinguishes between individual instances of the same type of object. For example, if the prompt text is to generate "picture of a garden with tulips, roses, and daffodils," instance segmentation would need to identify that the picture needs flowers of different types and generate the image for each flower separately.

Panoptic segmentation combines both semantic and instance segmentation and guides the model to generate or modify an image based on text input where the model recognizes, differentiates, and visually represents both "thing" classes (countable objects like animals, vehicles, furniture) and "stuff" classes (uncountable regions like grass, sky, water) in a single coherent image.

Stable Diffusion employs a U-Net network in the backward diffusion process to de-noise the image. The U-Net architecture is characterized by a symmetric "U" shape, which includes a contracting path to capture global context (this is equivalent to an encoder network) and an expansive or decoding path that enables precise local information collection. The output of the last layer in the expansive path is usually passed through a 1×1 convolution to map the feature vector to the desired number of output classes or dimensions.

Linking the encoder and the decoder networks is the skip connections. They connect the feature maps from the contracting path to the expansive path. Skip connections help in transferring fine-grained details and context information, which is essential for precise localization in tasks like segmentation or for detailed feature reconstruction in de-noising.

The architecture is often modified to include additional inputs, like text embeddings in text-to-image models, allowing the U-Net to condition its predictions on external information. A diagram of a U-Net network is shown in Figure 6-8.

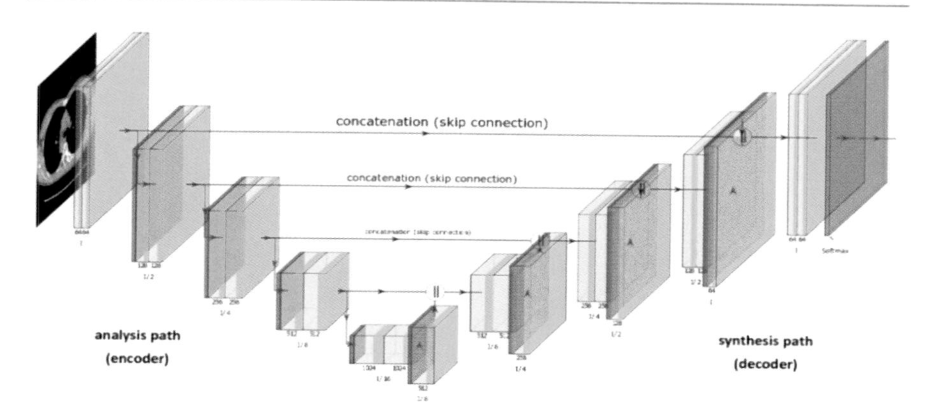

Figure 6-8 Diagram of a typical U-Net architecture. Note the use of skip connections between the encoder and the decoder networks [9]

Reinforcement Learning 7

Reinforcement learning (RL) is a type of machine learning where an agent learns to make decisions by performing actions in an environment that maximized the reward. The learning process involves the agent interacting with the environment, receiving feedback in terms of rewards or penalties, and using this feedback to refine its decision-making process. Gymnasium (formerly known as Gym), developed by OpenAI, is a popular toolkit for developing and comparing reinforcement learning algorithms. It provides a variety of environments ranging from simple toy tasks to complex real-world problems.

RL is a very useful tool in machine learning, but it does involve understanding a fair bit of mathematics to comprehend why the model chooses certain actions. If the reader is not familiar with Q-learning and deep network Q-learning, then I suggest the reader studies relevant literature before proceeding. Here, we will only discuss the RL at a high level, so the reader can follow the code.

7.1 Explanations of Reinforcement Learning

Reinforcement learning primarily revolves around the concepts of Markov decision processes (MDPs), providing a formal framework for decision-making where outcomes are partly random and partly under the agent's control. RL algorithms seek to learn "optimal" policies in this context. The optimal policy, in this context, maximizes the average of future rewards.

Imagine trying to deal with a discrete state problem where we are trying to act in an optimal way to maximize future rewards. Discrete state means considering discrete time steps and a manageable number of outcomes at any time step. These two assumptions are necessary to ensure that the problem is solvable. We further assume that future states depend only on the current state, known as the Markov process's memoryless assumption.

P. Hua, *Neural Networks with TensorFlow and Keras*,
https://doi.org/10.1007/979-8-8688-1020-6_7

$$Q^{\pi}(s,a) = R_t + E\gamma max_{a'}Q^{\pi}(S_{t+1},a'|S_t = s,a)$$

Equation 7-1 Bellman's equation in general form

The main point of these assumptions is that the optimal policy for reinforcement learning is specific to the Markov process. If we were to follow the same strategy for our daily life activities, such as making future investments, the optimal strategy would not be one suggested by Bellman's equation. Equation 7-1 states that the optimal value Q, following a policy π in state S with action a, is the immediate reward after taking action a plus the discounted sum of future rewards, which needs to be estimated. There are several ways to estimate this, including storing a very large table for every state-action pair S and A or run simulation.

For deep neural network learning, DQN uses a deep neural network to approximate the Q-value function in Equation 7-1. This network takes the state as input and outputs Q-values for all possible actions.

The network is trained by minimizing the loss function defined as the difference between the current Q-value estimate and the target Q-value from Bellman's equation. The loss function often used is the mean squared error:

$$L(\theta) = E(R_t + \gamma \max_{a'} Q^{\pi}(S_{t+1}, a'|S_t = s, a; \theta) - Q(S_t, A_t; \theta))^2 \tag{7.1}$$

θ are the weights in the neural network. In a standard DQN setup, there are two neural networks: the main network and the target network. The main network estimates the Q-values, while the target network, which is a delayed copy of the main network, provides the target Q-values for the loss calculation. This separation helps mitigate the problem of moving targets in the training process.

However, to simplify our code, we only use one neural network both for training and targeting, so it will take longer to train and be less stable, but it will still work as intended.

7.2 Gymnasium Library

For the purposes of reinforcement learning, we will be using the Gymnasium library, provided by the Farama Foundation. The Farama Foundation is a nonprofit organization dedicated to advancing the field of reinforcement learning through promoting better standardization and open source tooling for both researchers and industry.

The Gymnasium project's API contains four key functions: make, reset, step, and render. At the core of Gymnasium is Env, a Python class representing a Markov decision process (MDP) from reinforcement learning theory. Within Gymnasium, environments are implemented as Env classes, along with wrappers, which provide helpful utilities to alter the environment without writing boilerplate code. For example, in our Space Invaders game, we use a wrapper code to stack screen frames and change the environment into grayscale for more optimized training.

Before we can use Gymnasium, we need to install it. As Gymnasium licensing is non-distributable, the reader would need to install the library themselves using the procedure below.

7.2.1 Installing Gymnasium

Gymnasium can be installed using pip as follows:

```
pip install gymnasium[atari]
pip install gymnasium[accept-rom-license]
AutoROM --install-dir /path/to/install
pip install ale-py
ale-import-roms --import-from-pkg /path/to/install
```

This procedure not only installs Space Invaders but other Atari game environments as well.

7.2.2 Gymnasium

OpenAI Gymnasium is a project that provides an API for all single-agent reinforcement learning environments, including implementations of common environments, such as cartpole, pendulum, mountain-car, mujoco, Atari, and more. In this book, we will use Gymnasium to teach a machine how to play Space Invaders. For readers who do not know the game, Space Invaders is a 1978 shoot 'em up arcade video game developed and released by Taito in Japan. Commonly considered as one of the most influential video games of all time, Space Invaders was the first fixed shooter and set the template for the genre, such as *Galaxy* and *Missile Command*. The goal is to defeat wave after wave of descending aliens with a horizontally moving laser to earn as many points as possible.

The Gymnasium API provides a convenient way to explore RL, creating an environment that renders the game on-screen for a specific game. In our case, it displays the Space Invaders game in a window.

The agent, represented by our laser gun at the bottom of the screen, acts (moving left and right, staying still, or firing) as shown in Figure 7-1 as dictated by our RL policy and gets rewarded by the environment based on the number of aliens it shoots down. Our agent observes the game, which is a digitized version of the screen, in the form of a tensor. Incidentally, instead of the RGB tensor, Gymnasium also offers a 128K representation of the Atari environment, which is a compact state encoding designed to streamline training. Some researchers prefer this representation for its efficiency during training, particularly in resource-constrained settings. However, we will use the RGB tensor because it provides a more general framework that can be easily extended to non-Atari games.

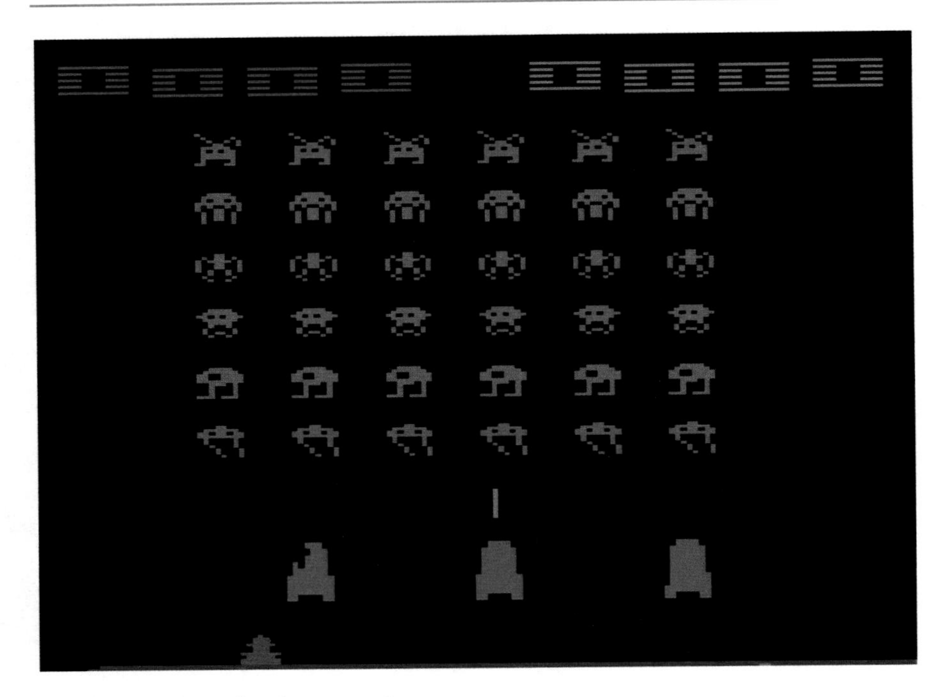

Figure 7-1 A Space Invaders screenshot

7.2.3 Explaining the Gymnasium Environment

Refer to Figure 7-2. This is a brief explanation of the Gymnasium environment for our game, but the environment for other games operates in the same manner.

First, an environment is created using make with an additional keyword "render_mode" that specifies how the environment should be visualized. See render for details on the default meaning of different render modes. In this example, we use the Space Invaders environment where the agent controls a laser to shoot aliens.

After initializing the environment, we reset it to obtain the first observation. At initialization, we have the option to start the environment with a particular random seed or options to reset the game.

Next, the agent performs an action in the environment, step; this can be imagined as moving a robot or pressing a button on a game's controller that causes a change within the environment. As a result, the agent receives a new observation from the updated environment along with a reward for taking the action. This reward could be, for instance, positive for destroying an enemy or a negative reward for moving into lava. One such action-observation exchange is referred to as a time step.

However, after some time steps, the environment may reach an end, known as the terminal state. For instance, if the number of lives has been exhausted, or the agent has succeeded in completing a task, the environment will need to stop as the

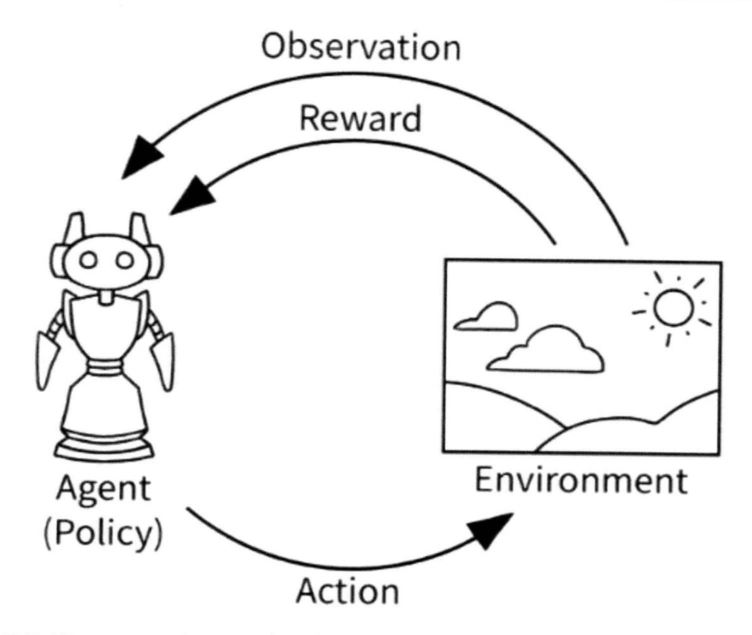

Figure 7-2 The agent-environment loop implemented in Gymnasium [10]

agent cannot continue. In Gymnasium, if the environment has terminated, this is returned by step. Similarly, we may also want the environment to end after a fixed number of time steps; in this case, the environment issues a truncated signal. If either of terminated or truncated is true, then reset should be called next to restart the environment.

Gymnasium Action and Observation Spaces

Each environment specifies the format of valid actions and observations with its action_space and observation_space attributes. This is helpful for both knowing the expected input and output of the environment as all valid actions and observation should be contained with the respective space.

To control the laser, we use an agent policy, mapping observations to one in the set of possible actions at each step.

Every environment should have the attributes action_space and observation_space, both of which should be instances of classes that inherit from space. Gymnasium has support for a majority of possible spaces users might need:

1. Box: Describes an n-dimensional continuous space. It is a bounded space where we can define the upper and lower limits which describe the valid values our observations can take.

2. Discrete: Describes a discrete space where 0, 1, . . ., $n - 1$ are the possible values our observation or action can take. Values can be shifted to $a, a + 1, \ldots, a + n - 1$ using an optional argument.

Preprocessing

The code for creating the environment is shown below, where we perform some preprocessing on the standard environment to optimize the learning process. The initial environment is defined as

```
env = gym.make("ALE/SpaceInvaders-v5",render_mode="rgb_array",
repeat_action_probability=0.1,
frameskip=1, obs_type='rgb')
```

- gym.make("ALE/SpaceInvaders-v5"): This command creates an instance of the Space Invaders game.
- render_mode="rgb_array": Sets the mode of rendering to return the game state as an RGB array.
- repeat_action_probability=0.1: Sets the probability that an action will be repeated instead of executing a new action. The reason why we need to do this is to introduce the exploration instead of exploitation in the action. The environment will, on average, move the laser randomly every one in ten actions.
- frameskip=1: Number of frames to skip between each action. Here, it is set to 1, meaning no frame skipping.
- obs_type='rgb': Sets the observation type to RGB, meaning the agent observes the game state in RGB color.

After setting up the standard environment, we introduce some preprocessing using the Atari wrapper library. These settings are chosen to simplify and standardize the input for the learning algorithm, making it easier for the algorithm to process and learn from the game environment:

```
env = gym.wrappers.AtariPreprocessing(env=env, noop_max=30,
                                      frame_skip=4,screen_size=84,
                                      terminal_on_life_loss=False,
                                      grayscale_obs=True,
                                      grayscale_newaxis=False,
                                      scale_obs=False)
```

- env=env: Passes the original Space Invaders environment.
- noop_max=30: Maximum number of "no-operation" (do nothing) actions to be performed at the start of an episode.
- frame_skip=4: The number of frames to skip (or repeat the same action) for each step. Here, it is set to 4.
- screen_size=84: Resizes the game screen to an 84x84 square for the observations.
- terminal_on_life_loss=False: If set to True, the episode ends when the player loses a life. Here, it is False, so the episode continues until all three lives are lost.

- grayscale_obs=True: Converts the RGB observation to grayscale, reducing the complexity of the input.
- grayscale_newaxis=False: Indicates whether a new axis should be added to the grayscale observation (not needed here). Since grayscale by default has only one channel, instead of passing in a 3D tensor, (W,H,3), we just use a 2D tensor (W,H).
- scale_obs=False: Decides whether to scale observations. Here, it is set to False, so observations are not scaled.

The line

```
env = gym.wrappers.FrameStack(env=env, num_stack=4,
          lz4_compress=False)
```

uses the wrapper provided by Gym that stacks consecutive frames of the game environment. This technique is often used in training reinforcement learning agents for video games because it helps the agent to understand motion and the temporal dynamics of the environment. By stacking frames, the agent can infer motion from the sequence of images, which is particularly important in games where understanding the dynamics (like the trajectory of moving objects) is crucial for performance. In Space Invaders, knowing the direction and speed of the invaders and projectiles can significantly improve the agent's ability to make decisions.

Let's consider frame_skip=4 and num_stack=4 in more detail:

Time step 1: The agent makes a decision, and the corresponding frame (Frame 1) is recorded.

Time steps 2–4: The agent's action from Time step 1 is repeated. Frames from these time steps are not recorded (due to frame skipping).

Time step 5: The agent makes another decision, and this frame (Frame 5) is recorded.

Time steps 6–8: Similar to Time steps 2–4, the action from Time step 5 is repeated, and these frames are not recorded.

Time step 9: Another decision is made, and Frame 9 is recorded.

Now, if we consider the stacked observation at Time step 9, it will consist of Frames 1, 5, and 9 for sure. The fourth frame in the stack is Frame 13 from the next decision point. There is an overlap between these stacked frames because Frame 5 was influenced by the decision made at Frame 1, and Frame 9 was influenced by the decision at Frame 5, and so on.

This overlapping of frames due to the combination of frame skipping and stacking provides the agent with a sense of temporal continuity, helping it to understand the consequences of its actions over several time steps, despite not observing every single frame in the sequence.

7.2.4 The Agent

```
class DQNAgent:
    def __init__(self, state_shape, action_size):
```

```
self.state_shape = state_shape # (4, 84, 84)
        self.action_size = action_size
        self.memory = deque(maxlen=2000)
        self.gamma = 0.95 # discount rate
        self.epsilon = 1.0 # exploration rate
        self.epsilon_min = 0.01
        self.epsilon_decay = 0.995
        self.learning_rate = 0.001
        self.model = self._build_model()

    def _build_model(self):
        # Neural Net for Deep-Q learning Model
        model = Sequential()
        model.add(Conv2D(32, (8, 8), strides=(4, 4),
              activation='relu',
              input_shape=self.state_shape,
              data_format='channels_first'))
model.add(Conv2D(64, (4, 4), strides=(2, 2),
              activation='relu',
              data_format='channels_first'))
model.add(Conv2D(64, (3, 3), activation='relu',
                   data_format='channels_first'))
model.add(Flatten())
model.add(Dense(512, activation='relu'))
model.add(Dense(256, activation='relu'))
model.add(Dense(self.action_size,
              activation='linear'))
model.compile(loss='mse', optimizer=
            Adam(learning_rate=self.learning_rate))
return model

def remember(self, state, action, reward,
           next_state, done):
        self.memory.append((state, action, reward,
                next_state, done))

def act(self, state):
        if np.random.rand() <= self.epsilon:
            return random.randrange(self.action_size)
        act_values = self.model.predict(state,verbose=0)
        return np.argmax(act_values[0])

def replay(self, batch_size):
        minibatch = random.sample(self.memory, batch_size)
        for state, action, reward, next_state,
          done in minibatch:
            target = reward if done else reward +
               self.gamma *
                 np.amax(self.model.predict(next_state,
                 verbose=0)[0])
            target_f = self.model.predict(state,verbose=0)
            target_f[0][action] = target
            self.model.train_on_batch(state, target_f)
```

```
            if self.epsilon > self.epsilon_min:
                self.epsilon *= self.epsilon_decay

def load(self, name):
        self.model.load_weights(name)

def save(self, name):
        self.model.save_weights(name)
```

This Python class, DQNAgent, represents a Deep Q-Network (DQN) agent for reinforcement learning and is a typical implementation of a deep network agent.

The constructor initializes the agent with the shape of the state space and the number of possible actions, action_size. It sets up a memory buffer using deque for experience replay, which stores past experiences up to a maximum length, 2000 in this case. It also defines the discount factor gamma for future rewards, the initial exploration rate epsilon, its minimum value epsilon_min, and the decay rate epsilon_decay. These are all the parameters for an RL model.

The model itself is a convolutional neural network (CNN) model for the Q-value function. The network takes the state frame stack as input and outputs Q-values for each action. The output layer has a size equal to the number of actions, with linear activation since Q-values can be any real number.

The constructor also sets the learning rate for the neural network and builds the model.

Method Remember Stores experiences (state, action, reward, next state, and done flag) in the replay memory. This data is used for training the model.

Method Act Implements the ϵ-greedy policy for action selection. With probability ϵ, it selects a random action (exploration), and with probability 1-ϵ, it chooses the best action based on the model's predictions (exploitation).

Method Replay Samples a mini batch from the memory and performs a learning step. The method updates the Q-values using the Bellman equation. The agent uses experience replay to break the correlation between consecutive experiences and stabilize the learning process.

7.2.5 Memory Replay

It is worth understanding why we need a replay buffer as this is not part of the standard RL literature. However, it is often used in RL for games for several reasons:

1. Reducing the Correlation Between Consecutive Experiences: In a sequential learning process, consecutive experiences are often highly correlated. In the case of Space Invaders, consecutive frames are highly correlated with the preceding frames because of the movement of objects on the screen. This correlation can

lead to instability and inefficiency in the learning process, as the model might overfit to recent experiences and not learn effectively from a more diverse set of experiences. By sampling randomly from the replay buffer, the experiences used in training are more varied, reducing the correlation and improving learning stability.
2. More Efficient Use of Past Experiences: In standard online learning, an experience would contribute to learning only at the time it occurs. With a replay buffer, past experiences are stored and reused in multiple learning updates, allowing the agent to learn more efficiently from a limited set of experiences. This is particularly important in environments where obtaining new experiences can be costly or time-consuming.
3. Reducing Variance in Updates: Learning from a batch of experiences, as opposed to learning from a single experience at a time, can help reduce the variance in the learning updates. This batch learning approach tends to lead to smoother, more stable learning over time.
4. Improving Data Efficiency: In complex environments, it's crucial to make the most of each piece of data. Experience replay allows the agent to learn from each experience multiple times, which is more data efficient than learning once and discarding the experience.
5. Balancing the Distribution of Experiences: Certain experiences may be rare but significant (like critical mistakes or successful strategies). A replay buffer ensures that these experiences have a lasting impact and are considered over many learning updates, rather than being quickly overshadowed by more frequent but less important experiences.
6. Stabilizing Learning in Deep Learning Models: When using deep neural networks for function approximation (as in DQN), training on correlated data can lead to catastrophic forgetting or unstable convergence. Experience replay mitigates this by providing a more stable, diverse set of experiences for training.

The most challenging part to explain, particularly if the reader is unfamiliar with Q-learning, involves the following lines of code:

```
    target = reward if done else reward + self.gamma *
np.amax(self.model.predict(next_state,verbose=0)[0])
    target_f = self.model.predict(state,verbose=0)
    target_f[0][action] = target
```

This code is part of the Q-learning update rule in a DQN. It adjusts the model's predictions based on the new information gained from taking an action and observing the outcome (reward and next state). The goal is to make the model's predictions of Q-values as accurate as possible, enabling the agent to make better decisions about which actions to take in different states.

To calculate the target Q-value, we use

```
    target = reward if done else reward + self.gamma *
np.amax(self.model.predict(next_state, verbose=0)[0])
```

This line calculates the target Q-value for the current state-action pair. If the game is over, then *done* is True, and the target is simply the immediate reward. This is because there are no future rewards to consider if the episode has ended.

If done is False, the target is calculated using the Bellman equation:

$$reward + self.gamma * np.amax(...) :$$

Here, self.gamma is the discount factor. It discounts the value of future rewards, reflecting the idea that immediate rewards are more valuable than distant rewards. np.amax(self.model.predict(next_state, verbose=0)[0]): This predicts the Q-values for all possible actions in the next state (next_state) and takes the maximum. This represents the best possible future reward that can be obtained from the next state, according to the current model.

The line

```
target_f = self.model.predict(state, verbose=0)
```

predicts the Q-values for all actions given the current state using the model. It essentially asks, "Given the current state, what are the Q-values (predicted future rewards) for each possible action?" The target_f is an array containing the predicted Q-values for each action in the current state. The size of target_f is the size of the model's output layer which has been set to the number of possible actions.

Once an action has been taken

```
target_f[0][action] = target
```

updates the Q-value for the action that was actually taken. The Q-value for this action, which was originally predicted by the model, is replaced with the target calculated earlier, and the model is retrained again with the better estimated Q-value.

```
self.model.train_on_batch(state, target_f)
```

The call to self.model.train_on_batch(state, target_f) updates the neural network's weights to minimize the difference between its predicted Q-values and the target_f Q-values. This update is done using backpropagation and an optimization algorithm Adam defined in the model's compilation step previously.

GPU Utilization for RL

The code written is not particularly optimized for GPU. For reinforcement learning using DQN and ALE, it is known that the GPU will be underutilized, so in this particular instance, it is better to run the project using a fast CPU rather than relying on a GPU.

Using Pretrained Networks

8

For some applications, it is almost mandatory to use networks that have been specifically pretrained to handle special tasks, such as large language models (LLMs), which involve complex natural language chatbots that are impossible for an individual to train. These research networks are typically straightforward to use through their APIs. The key is to thoroughly read and familiarize ourselves with the API documentation. These documents provide detailed instructions on making requests to the API, including information about endpoints, parameters, and other crucial details.

8.1 GPT-4

At the time of writing, OpenAI's GPT-4 is a paid service that includes DALL-E and the speech recognition tool Whisper. We must open an account with OpenAI and create an API key. Once this is completed, it is relatively straightforward to install and send queries. Before GPT-4, it was necessary to use less sophisticated libraries to perform natural language processing tasks, such as sentiment analysis, spam filtering, etc., but with GPT-4, these libraries have effectively became obsolete overnight. We can install the API for Python using

```
pip install openai
```

After which, we can create a client to stream requests and get responses back.

```
from openai import OpenAI
client = OpenAI()

stream = client.chat.completions.create(
model="gpt-4",
messages=["role": "user", "content": "Hello AI world!"],
stream=True,
)
for chunk in stream:
```


P. Hua, *Neural Networks with TensorFlow and Keras*,
https://doi.org/10.1007/979-8-8688-1020-6_8

```
if chunk.choices[0].delta.content is not None:
print(chunk.choices[0].delta.content, end="")
```

Generating images with DALL-E from a prompt is also easy. DALL-E will return an array of URLs, with the number determined by the parameter $n = 1$ in the code below. DALL-E 2 offers only "standard" quality, while the E-3 version includes an "HD" option.

```
from openai import OpenAI
client = OpenAI()

response = client.images.generate(
model="dall-e-3",
prompt="a white siamese cat",
size="1024x1024",
quality="standard",
n=1,
)
image_url = response.data[0].url
```

Currently, only DALL-E 2 supports the creation of edited image versions through masking and generating image variations, as shown in Figures 8-1 and 8-2.

```
from openai import OpenAI
client = OpenAI()

response = client.images.edit((
model="dall-e-2",
image=open("sunlit_lounge.png", "rb"),
mask=open("mask.png", "rb"),
prompt="A sunlit indoor lounge area
        with a pool containing a flamingo",
n=1,
size="1024x1024"
)
image_url = response.data[0].url
```

To create variation of an image, we would need to upload our own image.

```
from openai import OpenAI
client = OpenAI()

response = client.images.create_variation(
image=open("image_edit_original.png", "rb"),
n=2,
size="1024x1024"
)

image_url = response.data[0].url
```

Figure 8-1 OpenAI image generation using DALL-E [11]

Figure 8-2 Another example of image generation using prompting [11]

8.1.1 Fine-Tuning ChatGPT

One of the most powerful and useful applications using ChatGPT is the ability to fine-tune the model with our own dataset. Before fine-tuning though, OpenAI suggests we should go through prompt engineering, prompt chaining (breaking complex tasks into multiple prompts), and API function calling first as fine-tuning does require time and effort and money.

As an example, consider the following code example to fine-tune English to Italian.

8.1.1.1 Prepare the Dataset

The dataset should be a collection of English prompts and their Italian translations. This could be in a CSV file with two columns: "english" and "italian."

Example dataset (italian_language_dataset.csv):

```
english,italian
"Hello, how are you?","Ciao, come stai?"
"I am learning Italian.","Sto imparando l'italiano."
"What is your name?","Come ti chiami?"
...
```

Then write a Python script to do the training:

```
import openai
import pandas as pd
from sklearn.model_selection import train_test_split

# Load our dataset
df = pd.read_csv('italian_language_dataset.csv')

# Split the dataset into training and validation sets
train_df, valid_df = train_test_split(df, test_size=0.1)

def format_data(row):
return {"prompt": row['english'] + "\ n",
    "completion": row['italian'] + "\ n"}

# Format the data for OpenAI
train_data = list(train_df.apply(format_data, axis=1))
valid_data = list(valid_df.apply(format_data, axis=1))

# Set our API key
openai.api_key = 'your-api-key'

# Start the fine-tuning process
fine_tuning_response = openai.FineTune.create(
training_file=train_data,
validation_file=valid_data,
model="gpt-4",  # specify the model version
n_epochs=3,     # number of training epochs
batch_size=4,   # training batch size
)

print(fine_tuning_response)
```

The number of epochs depends on the complexity of the dataset in the same manner as normal training. If the task is very different from what the model was originally trained on, or if it is particularly complex, more epochs might be necessary.

Monitor the model's performance on a validation set and stop the training when the performance on this set starts to degrade. In practice, a common approach is to start with a relatively small number of epochs (like 3–5) and then adjust based on the model's performance and whether it shows signs of overfitting or underfitting. For fine-tuning large models like GPT-4, it is especially important to monitor performance closely, as these models can quickly overfit to a small dataset.

8.2 VGG

There are two common VGG models, VGG-16 and VGG-19. VGG stands for Visual Geometry Group; it is a standard deep convolutional neural network (CNN) architecture with multiple layers. The numbers 16 and 19 refer to the number of layers with VGG-16 or VGG-19 consisting of 16 and 19 convolutional layers, respectively.

Both models are used for image recognition. The VGGNet-16 supports 16 layers and can classify images into 1000 object categories, including keyboard, animals, pencil, mouse, etc. Both models take an input as an RGB image of size 224 × 224. The network architecture and an example of the feature retention through the layers are shown in Figure 8-3. Notice that as we go through the layers, only important and abstract features are retained as shown in Figure 8-4. We can either use VGGNet pretrained weights to classify objects, fine-tune the weights, or use a VGGNet pretrained layer to help with training our network as we did for the GAN project using VGG-19.

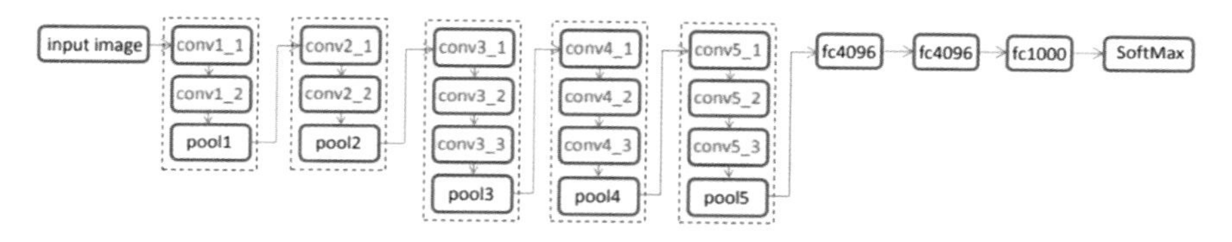

Figure 8-3 An example of feature retention for VGG-16

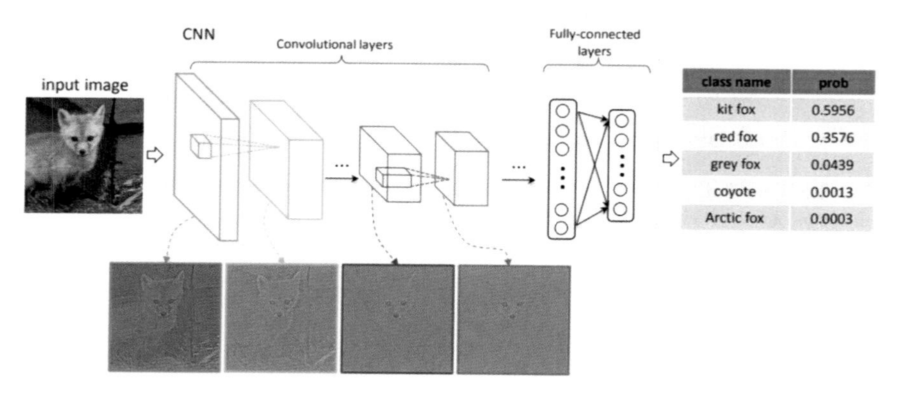

Figure 8-4 Network architecture for VGG-16 [12]

8.3 YOLO

YOLO, which stands for "You Only Look Once," is a popular series of models used for object detection in computer vision. YOLO is renowned for its speed and accuracy, making it a preferred choice for real-time object detection systems. It represents a significant shift from the traditional two-step approach of object detection (where one step identifies objects and another classifies them) to a single-step process.

Using YOLO is similar to other pretrained networks, and the following template should be familiar to readers:

```
import cv2
import numpy as np
import tensorflow as tf
from tensorflow.keras.models import load_model

# Load YOLO model assuming you
# have converted YOLO weights
# to a Keras model
yolo_model = load_model('yolo_model.h5')

# Function to process the image and return model's input
def process_image(image_path):
    image = cv2.imread(image_path)
    image = cv2.resize(image, (416, 416))
    image = image / 255.0  # Normalize pixel values
    image = image[np.newaxis, ...]  # Add batch dimension
    return image

# Process an image
input_image = process_image('path_to_your_image.jpg')

# Run the model
output = yolo_model.predict(input_image)

# Output needs to be post-processed to get
# bounding boxes and class labels
# This step can be complex and depends
# on the output format of your YOLO model
# Usually involves thresholding,
# non-max suppression, etc.

# For demonstration, let's assume we have
# a function to interpret the output
boxes, scores, classes = process_yolo_output(output,
                         threshold=0.5)

# Function to draw the results (boxes, scores, classes)
# on the image
# This is a simplified version, actual
# implementation may vary
def draw_boxes(image, boxes, scores, classes):
```

```
    for box, score, class_id in zip(boxes, scores, classes):
        x, y, w, h = box
        cv2.rectangle(image, (x, y), (x + w, y + h),
         (255, 0, 0), 2)
        cv2.putText(image, f"{class_id} {score:.2f}",
         (x, y - 5), cv2.FONT_HERSHEY_SIMPLEX,
            0.5, (255, 0, 0), 2)
    return image

# Draw the boxes on the original image and display/save it
output_image = draw_boxes(cv2.imread('path_to_your_image.jpg'),
     boxes, scores, classes)
cv2.imshow("Detection Output", output_image)
cv2.waitKey(0)
cv2.destroyAllWindows()
```

The only difference to note here is the conversion between YOLO and Keras. YOLO is implemented in Darknet which has a different framework structure than Keras.

8.3.1 Converting YOLO Weights to Keras

- Obtain YOLO Weights and Config Files: First, we need the original YOLO weights and configuration files. These are usually available from the official YOLO website or the repository of the specific YOLO version we are using.
- Choose a Conversion Tool: There are tools and scripts available online that can convert YOLO weights (which are typically in a format specific to the Darknet framework) to a format compatible with Keras. These tools read the Darknet configuration and weights files and then build a corresponding model in Keras, transferring the weights.
- Run the Conversion: Using the chosen tool, we can execute the conversion process. This usually involves specifying the paths to the input Darknet files and the desired output path for the Keras model file. The output is typically a .h5 file that contains the architecture and weights of the Keras model.
- Verification: After conversion, it is important to verify that the model performs as expected. This can be done by running the model on some test images and comparing the results with those obtained from the original YOLO model.

An example of the conversion using YAD2K (Yet Another Darknet 2 Keras) is shown below:

```
git clone https://github.com/allanzelener/YAD2K.git
cd YAD2K
python yad2k.py <yolo.cfg> <yolo.weights> <output_path.h5>
```

This script takes the YOLO config file (yolo.cfg) and the weights file (yolo.weights) and produces a .h5 file that can be loaded in Keras.

8.4 Hugging Face

We conclude this chapter with a summary of Hugging Face. This is not technically a pretrained network but is an open source library with a comprehensive collection of pretrained models, including NLP, RL, and CNN.

One of the key strengths of Hugging Face's library is its user-friendly interface. It allows for easy integration of complex models into applications, which has lowered the barrier to entry for working with advanced NLP models.

The library is too extensive to be fully covered in this book, but for readers interested in further ML studies, Hugging Face's tools and resources are extensively used in both academic and industry settings, offering access to state-of-the-art NLP technology.

8.5 Prompt Engineering

Up to now, we have been discussing the use of fine-tuning to tailor the model for our own needs. Fine-tuning involves adjusting the pretrained weights of a model on a specific dataset to improve its performance on tasks related to that dataset. This process customizes the model to better understand and generate outcome that aligns with the nuances of the new data. Fine-tuning is inflexible in the sense that each fine-tuned model is tailored to a particular task or domain. Changing the task might require retraining the model with different data. It also requires significant computational resources and expertise to retrain the model, especially for large language models. This process can be time-consuming and expensive.

Instead of fine-tuning, prompt engineering focuses on effectively interacting with language models, like ChatGPT, through the design of textual prompts. It encompasses strategies and techniques for crafting questions or instructions that guide these LLMs to generate desired outputs or perform tasks more accurately. The core of prompt engineering lies in understanding how a language model processes and responds to input, enabling users to achieve better results, whether it be generating text, coding, creating art, or solving problems. Prompt engineering can be performed by end users with an understanding of the model's capabilities and how it interprets language. However, in practice, there may be preprocessing processes which require coding as we shall now discuss.

We start with some prompting methods which are easily implemented. The users generate the relevant text prompts to guide the models toward a better result without the use of any preprocessing of data.

8.5.1 Zero-Shot Learning

In zero-shot learning, the model is given a task without any prior examples of how to perform it. The prompt must be self-explanatory, relying solely on the model's pretrained knowledge. This method is useful for general queries or tasks that do not require specialized knowledge beyond what the model was trained on. For example, suppose we want to translate a sentence from English to French. We would simply input the task instruction ("Translate the following sentence into French:") followed by the sentence. The model uses its preexisting knowledge from training to perform the translation. Zero-shot learning is powerful for tasks where the model's training data likely covered similar ground, and the task can be clearly articulated without examples.

8.5.2 Few-Shot Learning

Few-shot learning involves providing the model with a small number of examples (shots) to demonstrate how to perform a task. These examples are included directly in the prompt, giving the model a context to infer the task's requirements. Few-shot learning is ideal for tasks where the model needs a bit of guidance on the task's format or expected output but does not require extensive training data. In the translation example, for few-shot learning, we would provide the model with a few examples of English-to-French sentences and give the textual prompt ("Given these English-to-French translations, translate this new sentence:") and apply the pattern to the new input. Few-shot learning is useful when a bit of context can significantly improve the model's output by showing it exactly what is expected.

8.5.3 One-Shot Learning

One-shot learning is similar to few-shot learning except that we are giving the model a single example and ask it to generate similar results. Suppose we want the model to write jokes in a particular style. Here, the prompt would include a setup and punchline of a joke, followed by the task ("Write a joke in a similar style:"). The model uses this one example to understand both the structure and tone expected in its response. One-shot learning works well for tasks where a single example can clearly convey the task requirements and expected format.

8.5.4 Chain-of-Thought Prompting

Chain-of-thought prompting encourages the model to break down complex problems into smaller, more manageable steps. This structured approach to problem-solving is akin to how humans tackle difficult tasks, making it easier for the model to navigate through the complexities of a problem and arrive at a solution more systematically.

LLMs are trained on vast amounts of text data, encompassing a wide range of topics and problem-solving strategies. Chain-of-thought prompting leverages this preexisting knowledge by guiding the model to apply relevant information and reasoning patterns it has learned during training to the task at hand.

By requiring the model to articulate each step in its reasoning process, chain-of-thought prompting makes the model's thought process more transparent. This not only helps in understanding how the model arrived at a particular conclusion but also aids in identifying and correcting errors in the model's reasoning. With each step of the thought process laid out, it becomes easier to identify where the model might have gone off track. This stepwise breakdown allows for targeted adjustments to the prompt or the reasoning process, facilitating more accurate outcomes.

In some areas where LLMs are weak currently, such as mathematically reasoning and coding, guiding the model in several steps often yields more accurate and better results. In some cases, however, the model is unable to proceed any further and will provide the same incorrect answer but written in a different way. In such cases, it is necessary to tackle the problem differently by asking questions with different topics, which hopefully is addressable by the trained model.

For example, instead of prompting the model to solve $2x + 3 = 7$ directly, we ask it to "Explain step by step how to solve $2x + 3 = 7$." The model then breaks down the problem into a series of logical steps leading to the solution. This method is ideal for complex tasks requiring transparency in the problem-solving process, helping users follow the model's reasoning and ensuring the solution's correctness.

8.5.5 Role-Playing

Role-playing prompting is a creative and effective approach in prompt engineering, especially useful when interacting with large language models (LLMs), like GPT-3 or GPT-4. This technique involves asking the model to assume a specific character, identity, or perspective when generating its responses. By doing so, it influences not only the content of the model's output but also its tone, style, and the type of information it prioritizes.

A role-playing prompt would explicitly state the role ("Imagine you are a science teacher explaining the theory of relativity to a 10-year-old. How would you describe it?"). This guides the model to adopt a specific tone, complexity level, and perspective, tailoring the response to fit the role. Role-playing enhances creativity and engagement, making it suitable for educational content, entertainment, and more. This prompt encourages the model to adopt a role (a science teacher) and a specific tone (friendly and enthusiastic) tailored to the audience (a 10-year-old child). The model's response is likely to be more engaging, understandable, and appropriate for a young audience than a straightforward factual explanation.

8.5.6 Embedding Prompts

The prompting methods up to now do not require coding. Unlike these methods, embedding prompting needs coding and a deeper understanding of how the LLMs work.

Embedding prompts in the context of working with language models involve using vector representations of text (embeddings) to guide or influence the model's responses, rather than relying solely on traditional text-based prompts. This approach can be particularly powerful for tailoring model outputs, improving relevance, and achieving more nuanced interactions. We have seen how embedding is used in the Stable Diffusion model with the data being preprocessed into a latent space vector first instead of being used directly. By compressing text prompts into dense vector representations of words, phrases, or longer texts, we are able to help the model to capture the semantic meaning in a high-dimensional space. The process of embedding enables the model to capture nuanced relationships between different pieces of text.

Here is a closer look at how embedding prompts work and some examples:

- Semantic Search: Suppose we are building a system to find relevant documents based on a query. Instead of matching keywords, we use embeddings to represent the query and the documents. The system then retrieves documents whose embeddings are closest to the query's embedding, likely resulting in more relevant matches. For example, for a query "sustainable energy sources," the system finds documents related to solar power, wind energy, etc., even if the exact phrase is not used.
- Content Recommendation: In a recommendation engine, user preferences and item descriptions are represented as embeddings. The system recommends items by finding those whose embeddings are closest to the user's preference embedding. For example, if a user likes articles about "space exploration," their preference embedding might closely align with articles about Mars rovers or satellite launches, and those articles would be recommended.
- Dialogue Systems: For generating contextually relevant responses in a chatbot, the embeddings of the conversation context and potential responses are used. The chatbot selects or generates a response whose embedding is most aligned with the context embedding, ensuring relevance and coherence in the conversation. For instance, if the conversation context involves discussing "favorite books," the chatbot focuses its responses on literature-related content.

8.5.7 Knowledge Graphs

It may seem counterintuitive that a knowledge graph can be fed into an LLM to guide the output, but it does work quite effectively.

A knowledge graph is a type of graphical representation that uses nodes (vertices) to represent entities or concepts and edges to represent the relationships between them. These graphs can be directed or undirected and can include various types of relationships, such as hierarchical (parent-child), associative (related or connected ideas), or sequential (steps in a process). Relationship graphs, when used in the context of prompting language models, provide a structured way to represent and utilize the relationships between entities, concepts, or topics to guide the generation of text. This approach can significantly enhance the relevance, coherence, and depth of the responses generated by AI models. Some applications of knowledge graphs include

- Educational Content Creation: When generating content on a historical event, a relationship graph could map out key figures, locations, causes, effects, and timelines. The prompt could guide the model to elaborate on these elements based on their connections, ensuring comprehensive coverage.
- Storytelling: In creative writing, a relationship graph could outline characters, settings, plot points, and thematic elements. The model could then be prompted to weave a narrative that respects these relationships, enhancing plot coherence and character development.
- Technical Documentation: For generating technical or scientific content, a graph could represent concepts, principles, applications, and examples. Prompts could leverage this structure to produce detailed explanations that accurately reflect the complexity and hierarchy of technical knowledge.

The current model of GPT-4 and similar AI models from OpenAI do not inherently "understand" visual data, including spider graphs, in the way humans do. These models are primarily designed to process and generate text-based information. However, GPT-4 can interpret descriptions of spider graphs (or any other visual information) provided in text format. This means that while the model cannot directly analyze a spider graph image or its visual components, it can understand and respond to detailed textual descriptions of the data or insights that the spider graph is intended to convey.

Suppose we have a spider graph comparing five different products across multiple criteria: price, quality, user interface, customer support, and feature set. We cannot show GPT-4 the graph directly but can describe it as follows:

"Product A scores high on price and quality but low on user interface and customer support. Product B has moderate scores across all criteria, with a slight edge in feature set. Products C and D excel in user interface and customer support, respectively, but fall short in quality. Product E is balanced but does not lead in any criterion."

Given this description, we could ask GPT-4 questions like

"Which product offers the best balance across all criteria?" "Based on the description, which areas should Product A focus on improving?" "How might improving customer support impact the overall score of Product D?"

8.6 Retrieval-Augmented LLM

One major problem of LLMs is the so-called knowledge cutoff. The system may have been trained on out-of-date data, or the data may be classified so that it does not have any context-dependent knowledge on the problem at hand. The latest version of ChatGPT automatically searches the Internet if it cannot find the relevant answer. In effect, it has changed into a retrieval-augmented LLM.

Retrieval-augmented large language models (LLMs) are a class of AI that enhance traditional LLMs by integrating external data retrieval into the response generation process. This approach allows the models to access and incorporate real-time information or domain-specific knowledge from external databases or the Internet, thereby extending the model's knowledge base beyond its initial training data. This augmentation significantly improves the model's ability to provide accurate, up-to-date, and contextually relevant answers, making it particularly useful for tasks requiring current knowledge or specialized information across various domains.

To use a retrieval-augmented LLM for our own specific information, we would integrate the LLM with a system capable of querying our company's databases, knowledge bases, or the Internet for real-time information. The LLM can then incorporate this fetched data into its generated text, providing accurate and up-to-date responses based on your company's latest information. This could enhance customer support chatbots, automate content creation for marketing, or support decision-making by providing insights from the latest data. Implementing such a system requires technical expertise in AI, databases, and possibly API development for seamless integration.

8.7 Best Practices for Prompt Engineering

These are the best prompt formats recommended by OpenAI taken from their website as of March 2024:

- Put instructions at the beginning of the prompt and use ### or """ to separate the instruction and context.
  Less effective ✗:
  Summarize the text below as a bullet point list of the most important points.
  text input here
  Better ✓:
  Summarize the text below as a bullet point list of the most important points.
  Text: """ text input here """
- Be specific, descriptive, and as detailed as possible about the desired context, outcome, length, format, style, etc.
  Less effective ✗:
  Write a poem about OpenAI.
  Better ✓:
  Write a short inspiring poem about OpenAI, focusing on the recent DALL-E product launch (DALL-E is a text to image ML model) in the style of a famous poet.
- Articulate the desired output format through examples.
  Less effective ✗:

Extract the entities mentioned in the text below. Extract the following 4 entity types: company names, people names, specific topics, and themes.
Text: text
Show and tell—the models respond better when shown specific format requirements. This also makes it easier to programmatically parse out multiple outputs reliably.
Better ✓:
Extract the important entities mentioned in the text below. First extract all company names, then extract all people names, then extract specific topics which fit the content, and finally extract general overarching themes.
Desired format: Company names: <comma-separated list of company names> People names: -||- Specific topics: -||- General themes: -||-
Text: text

- Start with zero-shot, then few-shot, neither of them worked, then fine-tune.
  ✓Zero-shot
  Extract keywords from the below text.
  Text: text
  Keywords:
  ✓Few-shot: Provide a couple of examples.
  Extract keywords from the corresponding texts below.
  Text 1: Stripe provides APIs that web developers can use to integrate payment processing into their websites and mobile applications. Keywords 1: Stripe, payment processing, APIs, web developers, websites, mobile applications ##. Text 2: OpenAI has trained cutting-edge language models that are very good at understanding and generating text. Our API provides access to these models and can be used to solve virtually any task that involves processing language. Keywords 2: OpenAI, language models, text processing, API. ## Text 3: text Keywords 3:
  ✓Fine-tune: See fine-tune best practices; refer to their documentation at https://platform.openai.com/docs/guides/fine-tuning.
- Reduce "fluffy" and imprecise descriptions. Less effective ✗:
  The description for this product should be fairly short, a few sentences only, and not too much more.
  Better ✓:
  Use a 3–5 sentence paragraph to describe this product.
- Instead of just saying what not to do, say what to do instead.
  Less effective ✗:
  The following is a conversation between an agent and a customer. DO NOT ASK USERNAME OR PASSWORD. DO NOT REPEAT.
  Customer: I can't log in to my account. Agent:
  Better ✓:
  The following is a conversation between an agent and a customer. The agent will attempt to diagnose the problem and suggest a solution while refraining from asking any questions related to PII. Instead of asking for PII, such as username or password, refer the user to the help article: www.samplewebsite.com/help/faq.

Customer: I can't log in to my account. Agent:

- Code Generation Specific: Use "leading words" to nudge the model toward a particular pattern.
  Less effective ✗:
  # Write a simple python function that # 1. Ask me for a number in mile # 2. It converts miles to kilometers
  In this code example below, adding "import" hints to the model that it should start writing in Python. (Similarly, "SELECT" is a good hint for the start of a SQL statement.)
  Better ✓:
  # Write a simple python function that # 1. Ask me for a number in mile # 2. It converts miles to kilometers.
  import

8.7.1 Parameters

Generally, we find that model and temperature are the most commonly used parameters to alter the model output.
model: Higher performance models are generally more expensive and may have higher latency.
temperature: A measure of how often the model outputs a less likely token. The higher the temperature, the more random (and usually creative) the output. This, however, is not the same as "truthfulness." For most factual use cases, such as data extraction and truthful Q&A, the temperature of 0 is best.
max_tokens (maximum length): Does not control the length of the output, but a hard cutoff limit for token generation. Ideally, we will not hit this limit often, as our model will stop either when it thinks it is finished or when it hits a stop sequence we defined.
stop (stop sequences): A set of characters (tokens) that, when generated, will cause the text generation to stop.

8.8 Coding an AI Agent Using LangChain

One useful application for LLMs is the ability to create intelligent and autonomous AI agents. An AI agent refers to a system or software that is capable of acting autonomously in an environment to meet its designed objectives or goals. These agents can make decisions and perform actions without human intervention, based on the data they receive and their preprogrammed rules or learning algorithms. AI agents are capable of observing their environment through sensors and acting upon that environment with actuators.

The capabilities of AI agents can range from simple, rule-based systems to complex, learning-based systems that utilize machine learning, deep learning, or reinforcement learning to adapt and improve their performance over time. AI agents are used in a wide array of applications, including virtual personal assistants, autonomous vehicles, smart home devices, game playing, financial trading, healthcare for diagnosis and treatment recommendations, and industrial automation for optimizing processes.

We will make use of LangChain to create a simple AI agent. LangChain is a powerful framework for quickly prototyping LLM applications by chaining together LLM tasks and running autonomous agents quickly and easily. LangChain consists of four main components:

- LangChain Libraries: The Python and JavaScript libraries. Contain interfaces and integrations for a myriad of components, a basic runtime for combining these components into chains and agents, and off-the-shelf implementations of chains and agents.
- LangChain Templates: A collection of easily deployable reference architectures for a wide variety of tasks.
- LangServe: A library for deploying LangChain chains as a REST API.
- LangSmith: A developer platform that lets you debug, test, evaluate, and monitor chains built on any LLM framework and seamlessly integrates with LangChain.

For our AI agent, we will implement code to augment OpenAI with information from our database. The process of bringing the appropriate information and inserting it into the model prompt is known as retrieval-augmented generation (RAG).

LangChain has several tools to deal with questions and answers (Q&A) for unstructured data using RAG. One such tool is Q&A over SQL data. We will use this tool to build a simple application which embeds and stores data and then retrieves and generates answers using OpenAI LLM queries. The embedding and storing are done offline, while the data retrieval and answer generation are carried out in real time.

Our RAG architecture is simple and is typical of a RAG application. It has two components:

- Indexing: A pipeline for ingesting data from a source and indexing it. This usually happens offline.
- Retrieval and Generation: The actual RAG chain, which takes the user query at runtime and retrieves the relevant data from the index, then passes that to the model.

8.8.1 Indexing Using VectorDB

For the indexing part, there are three steps involved as depicted in Figure 8-5. The data loading step loads data from a variety of sources, including PDFs, text, URLs, and images, and is done using the *DocumentLoaders* toolkit.

The document loader has both built-in functions for loading text, CSV, PDF, image, JSON, and markup as well as third-party community libraries for loading

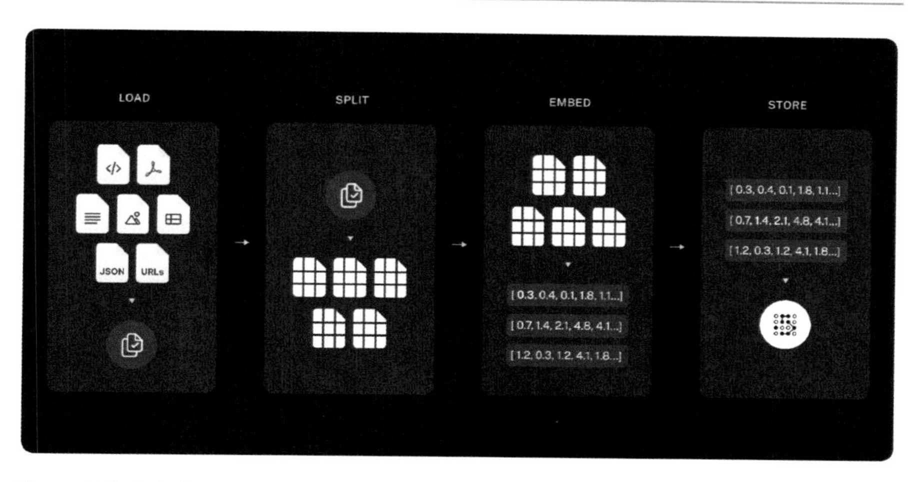

Figure 8-5 Indexing procedure in RAG [13]

other types of data and websites, such as Wikipedia, WhatsApp chat, and YouTube.

The Split toolkit is a text splitter which breaks large *Documents* into smaller chunks. This is useful both for indexing data and for passing it in to a model, since large chunks are harder to search over and will not fit in a model's finite context window.

The *Vector Store* and *Embeddings* models are responsible for embedding and storing vectors. These models transform data into a latent format specific to a particular large language model (LLM). For example, when using the ChatGPT model, data embedding is performed using the ChatGPT-specific embedding technique. Conversely, if the Gemini model is employed, data must be embedded using the embedding method unique to Gemini. Once the data is embedded, when we query the data at query time, we embed the unstructured query and retrieve the embedding vectors that are "most similar" to the embedded query. A diagrammatic view of storing and querying a vector store is shown in Figure 8-6. The LangChain Indexing API syncs your data from any source into a vector store, helping us to avoid writing duplicated content into the vector store, rewriting unchanged content, and recomputing embeddings over unchanged content.

A vector store takes care of storing embedded data and performing vector search for us. The vector store itself does not impose an embedding method. A vector store is simply a specialized database or storage system designed to efficiently store and manage vector embeddings. Vector embeddings, as we have seen, are high-dimensional representations of data, typically text, images, or any form of content that has been converted into numerical vectors using various embedding techniques. These embeddings capture the semantic or contextual similarities between data points in a way that can be easily processed by algorithms.

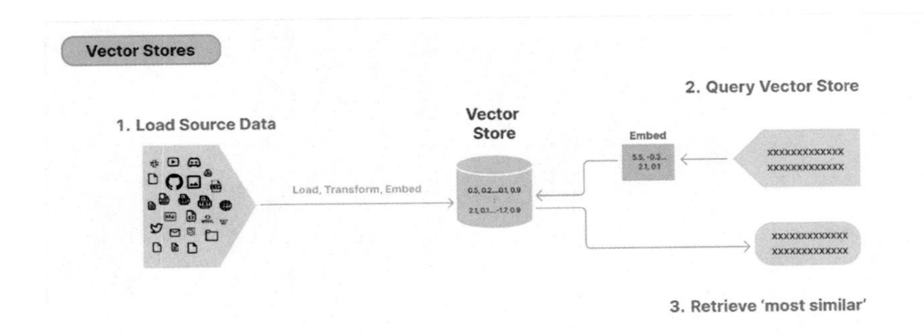

Figure 8-6 Querying a vector store [14]

Vector stores are optimized to handle the specific challenges associated with storing and querying high-dimensional data. They support operations such as nearest neighbor search, which is the task of finding the most similar vectors in the store to a given query vector. This is crucial in applications like recommendation systems, similarity searches, and clustering, where understanding the relationship between data points based on their embeddings is essential.

There are many open source vector stores available which can be installed and run locally. For our example, we choose Chroma VectorDB because it is lightweight and easy to configure, although the FAISS vector database, which makes use of the Facebook AI Similarity Search (FAISS) library, is another possible candidate.[1]

8.8.2 Retrieval Mechanism in LangChain

Once data is stored in the document storage, which in general could be a data store other than VectorDB, retrieving it effectively is crucial. LangChain offers several retrieval algorithms, such as basic semantic search. Beyond this basic approach, LangChain enhances performance with an array of advanced algorithms via its retrievers, including

- Parent Document Retriever: Enhances retrieval flexibility by generating multiple embeddings for each parent document. This feature enables queries on specific segments while providing results within a broader context.
- Self-Query Retriever: Recognizes that user queries often entail more than semantic content, incorporating logical references that are best addressed through metadata filtering. This retriever separates the semantic elements from metadata filters within a query, optimizing the retrieval process.

[1] FAISS is powerful but challenging to configure for a particular application.

Table 8.1 Overview of retrieval methods [15]

Name	Index type	Uses LLM	When to use	Description
Vector store	Vector store	No	Getting started, simplicity preferred	The simplest method, ideal for beginners. Involves creating embeddings for each text piece
Parent document	Vector store + document store	No	Distinct information in documents	Indexes multiple chunks per document, retrieving the whole document for similar chunks
Multivector	Vector store + document store	Sometimes	Relevant information extraction	Creates multiple vectors per document for diverse indexing strategies
Self-query	Vector store	Yes	Metadata-based queries	Splits user queries into semantic searches and metadata filters
Contextual compression	Any	Sometimes	Reducing irrelevant information	Applies post-processing to extract the most relevant information from documents
Time-weighted vector store	Vector store	No	Document recency is important	Combines semantic similarity and timestamp data for retrieval
Multi-query retriever	Any	Yes	Complex queries	Generates multiple queries from one to cover various topics
Ensemble	Any	No	Combining retrieval methods	Uses multiple retrievers for comprehensive document fetching
Long-context reorder	Any	No	Attention to document context	Reorders documents to highlight the most relevant information at the beginning and end

- Ensemble Retriever: Designed for scenarios requiring document retrieval from varied sources or through multiple algorithms. This retriever streamlines the process, enabling efficient and comprehensive document retrieval across diverse databases and methodologies.

The website for LangChain provides a succinct table for the different retrieval methods and when to use them, which is displayed in Table 8.1.

We end this section on LangChain by providing snippets of code to demonstrate the various stages in building an in-house AI agent to help with learning LangChain. The source of data will be web pages from Langchain.com, instructional videos, and

PDFs. The downloaded data will be cleaned and stored in a VectorDB. We will then augment OpenAI with our database and send some pertinent queries to view the result.

8.9 Company Chatbot Using LangChain

First, we install all the necessary packages. In this exercise, we will use the OpenAI LLM, so you will need to subscribe to OpenAI and get the API key. In addition, at least at the time of writing, you will also have to pay OpenAI a small fee to query its database.

```
pip install langchain-openai
pip install beautifulsoup4
pip install unstructured
pip install lxml
```

The beautiful soup and unstructured packages are used to streamline and optimize the data processed around LLMs. The beautiful soup scrapes the web pages, while the unstructured library provides tools to preprocess text documents, PDFs, and HTMLs.

```
def scrape_page_and_save(url, output_dir,
      base_url, depth=0, max_depth=3):
if depth >= max_depth:
    # Terminate recursion if maximum depth is reached
    return
# Skip 'mailto' URLs
if url.startswith('mailto:') or not url.startswith(base_url):
    return
try:
    response = requests.get(url)
    soup = BeautifulSoup(response.text, 'lxml')
    paragraphs = soup.find_all('p')
    # Check if the page has any content
    if not paragraphs:
        return  # Skip empty pages
    filename = os.path.join(output_dir, url.
        split('/')[-1] + '.txt')
    with open(filename, 'w', encoding='utf-8') as f:
        for paragraph in paragraphs:
        line = paragraph.text.strip() + '\n'
        try:
            f.write(line)
            f.flush()
        except Exception as e:
            print(f"Error writing to file: {e}")
    # Find links to other pages and recursively scrape them
    links = soup.find_all('a', href=True)
    for link in links:
        absolute_url = urljoin(url, link['href'])
        # Recursively scrape the subpage and save it
```

```
            scrape_page_and_save(absolute_url,
                output_dir, base_url, depth=depth +
                1, max_depth=max_depth)
    except requests.exceptions.RequestException as e:
        print(f"Error occurred while scraping {url}: {e}")
```

The code above will recursively scrape all child pages and parent page. However, it does not scrape external pages to the base URL to prevent links to GitHub via the code listing within the child pages.

In addition to extracting data from web pages, a transcription of YouTube videos is occasionally required. There are two main methods for accomplishing this. The first method involves downloading the video via their API, followed by employing the OpenAI Whisper API to transcribe the text from the resulting audio file. Alternatively, if a transcription is already available, tools like the youtube-transcript-api can be utilized to convert the text into English. One drawback of using the Whisper API is its cost, as it is a paid service, whereas the youtube-transcript-api is freely available as open source software.

We will illustrate the transcription using both methods. We begin by installing the packages as usual:

```
pip install pytube moviepy openai-whisper
pip install youtube-transcript-api
```

The YouTube transcript API transcribes the text in a particular language and converts the data into a consistent string of a given format, such as a basic text (.txt), or even formats that have a defined specification, such as JSON (.json), WebVTT (.vtt), SRT (.srt), comma-separated format (.csv), etc.

To demonstrate its application, we will transcribe a tutorial video from LangChain on creating an open source software (OSS) model retrieval agent with the help of Mistral and Nomic Embed Text. Initially, we'll download the video and then use the youtube-transcript-api to transform it into text.

```
def text_from_YouTube_video(video_id,output_file):
    transcript = YouTubeTranscriptApi.
        get_transcript(video_id,languages=['en'])
    formatter = TextFormatter()
    text_format = formatter.format_transcript(transcript);
    # remove some characters from the transcript
    text_format = text_format.replace('\n', ' ').
        replace('um', '').replace('uh', '')
    with open(output_file, 'w', encoding='utf-8') as text_file:
        text_file.write(text_format)

youtube_url = 'Ce03oEotdPs'
text_from_YouTube_video(youtube_url,output_directory
    + "/langchainvideo.txt")
```

Often, numerous instances of "um" and "uh" are found, which are then eliminated by the .replace() line prior to saving to a text file.

As an alternative to using the YouTubeTranscriptApi, we can employ the Whisper API to transcribe the audio portion of the video. To transcribe a YouTube video using OpenAI's Whisper API in Python, we will start by downloading the video from YouTube. This can be achieved through the pytube library, which allows for the retrieval of videos via their URLs. After downloading the video, we may need to extract its audio component to ensure that the Whisper API can transcribe the spoken content accurately. The MoviePy tool can assist in converting the video file into an audio format, such as MP3, compatible with the Whisper API.

After saving the audio file, the subsequent step involves using OpenAI's Whisper model to transcribe the audio to text. The Whisper Python package provides a straightforward method for loading the model and processing the audio file to generate a transcription. By indicating the path to the audio file, the API processes the content and delivers the transcribed text. We transcribe the same video as before and save it down so that the reader can compare the accuracy between the two conversion methods.

```
def download_youtube_audio(youtube_url, filename):
    # Download YouTube video audio stream
    video = YouTube(youtube_url).streams.filter(only_audio=True).
        first()
    download_path = video.download()  # This downloads the audio
        in its native format (usually .mp4 or .webm)

    base, ext = os.path.splitext(download_path)
    os.rename(download_path, filename)

def transcribe_audio_with_whisper(filename,output_file):
    # Load the Whisper model
    model = whisper.load_model("base")

    # Transcribe the audio file
    result = model.transcribe(f"{filename}")
    text_format = result['text']
    text_format = text_format.replace('\n', ' ').replace('um',
        '').replace('Um', '').replace('Uh', '').replace('uh', '')
    with open(output_file, 'w', encoding='utf-8') as text_file:
    text_file.write(text_format)
```

Once the data downloading is completed, the next step is text preprocessing and feature extraction. Text data often comes with noise and formatting that may not be relevant for analysis. Preprocessing aims to clean and standardize the text, which might include

- Removing special characters and punctuation.
- Lowercasing all text to maintain consistency.
- Eliminating stopwords (common words that add little semantic value). For this task, we will use the Natural Language Toolkit (nltk) for stopword removal simply by using pip install nltk.
- Segmenting longer documents into smaller chunks to ensure that embeddings accurately represent the document's content.

A simple function to preprocess the data is shown below. In a large-scale project, we might use a more sophisticated library to handle this task. Commercial tools like IBM Watson Natural Language Understanding or open-source libraries such as spaCy are highly robust and capable of managing a wide variety of textual inputs, making them versatile for processing raw or minimally processed text. However, customized preprocessing can fine-tune the input data to better align with specific objectives, potentially enhancing the relevance and accuracy of the model's responses.

After preprocessing, we use OpenAI embedding and FAISS to store and index the tokens. Technologies like FAISS play a crucial role in this component by enabling fast retrieval of similar embeddings. If you have a local GPU, then install FAISS using pip install faiss-gpu; otherwise, use pip install faiss-cpu.

An embedding is a high-dimensional vector that represents text in a form that captures its semantic meaning. OpenAI provides an embeddings API that supports various models optimized for embedding generation. Using this API, each document or text snippet is transformed into a vector. We had to deal with a similar problem in our VAE model.

The rest of the code to create a query agent is described below:

```
import shutil
import os
from langchain.vectorstores.faiss import FAISS
from langchain.prompts import ChatPromptTemplate
from langchain.embeddings.openai import OpenAIEmbeddings
from langchain.document_loaders import DirectoryLoader,
    TextLoader
from langchain.text_splitter import
    RecursiveCharacterTextSplitter
from langchain.chat_models import ChatOpenAI

FAISS_PATH = "~/AI/Data/Chatbot/faiss_index"
llm = ChatOpenAI(api_key=os.environ['OPENAI_API_KEY'])
PROMPT_TEMPLATE = """
Use the following context to answer the question:
{context}
---
Answer the question based on the above context: {query}"""


def save_directory_text_to_store(path):
    text_loader_kwargs = {'autodetect_encoding': True}
    loader = DirectoryLoader(path, glob="**/*.txt", loader_cls=
        TextLoader, loader_kwargs=text_loader_kwargs,
    show_progress=True)
    documents = loader.load()
    text_splitter = RecursiveCharacterTextSplitter(chunk_size
        =2000, chunk_overlap=200, length_function=len,
    add_start_index=True)
    chunks = text_splitter.split_documents(documents)
    if os.path.exists(FAISS_PATH): shutil.rmtree(FAISS_PATH)
    db = FAISS.from_documents(chunks, OpenAIEmbeddings())
    db.save_local(FAISS_PATH)
```

```
    print(f"Save {len(chunks)}")
    return db

def get_db():
    embedding_function = OpenAIEmbeddings()
    db = FAISS.load_local(FAISS_PATH, embedding_function)
    return db
```

The function save_directory_text_to_store(path) loads text files from a specified directory(path), auto-detects their encodings, and reads their content. It then splits the documents into chunks using a RecursiveCharacterTextSplitter, which is configured to divide the text into segments with a specific chunk size and overlap, aiding in handling large documents and ensuring continuity of context across chunks. The reason why we need to overlap the text is to provide context for the model; otherwise, the individual chunks of text will be disjointed.

If the FAISS index exists at FAISS_PATH, it deletes the existing index to prevent duplication. It also uses the FAISS module to create a new vector store from the document chunks, with embeddings generated via OpenAIEmbeddings.

Finally, the function saves the newly created FAISS index to the local filesystem for later retrieval.

As the reader can see, LangChain has hidden most of the complexity in preprocessing the data into its tools so that the user can concentrate on high-level coding.

Once we have created a vector database and are able to query it, performing RAG with OpenAI can be done with the code below:

```
def query_openai_and_faiss(store, query_text):
    # use the faiss vector store we saved to search the local
        document
    retriever = store.as_retriever(search_type="mmr",
        search_kwargs={"k": 3})
    results = retriever.get_relevant_documents(query_text)

    context_text = "\n\n---\n\n".join([doc.page_content for doc
        in results])
    prompt_template = ChatPromptTemplate.from_template(
        PROMPT_TEMPLATE)
    prompt = prompt_template.format(context=context_text, query=
        query_text)
    print(prompt)
    response_text = llm.predict(prompt)
    sources = [doc.metadata.get("source", None) for doc in
        results]
    index = [doc.metadata.get("start_index", None) for doc in
        results]
    formatted_response = f"Response: {response_text}\nSources: {
        sources} index: {index}"
    print(formatted_response)
    return response_text
```

The program above outputs the following response to the query:

Answer the question based on the above context: How can I use LangChain to build a chatbot?

Response: To build a chatbot using LangChain, you can follow the steps outlined in the quickstart guide provided. First, you will need to install the LangChain Anthropic integration package and obtain an API key by creating an account. Then, you can set the API key as an environment variable or pass it directly when initializing the Anthropic Chat Model class.

After setting up the integration package, you can initialize the LLM of your choice and use prompt templates to guide the responses. Additionally, you can utilize retrieval to provide additional context to the LLM by fetching relevant data from external sources.

By following the instructions in the quickstart guide and utilizing features such as prompts, models, output parsers, and retrieval, you can successfully build a chatbot using LangChain. Remember to refer to the documentation for a deeper dive into the process and for more advanced techniques. Sources:

```
['~/AI/Data/Chatbot/get_started.txt',
'~/AI/Data/Chatbot/quickstart#updating-retrieval.txt'
'~/AI/Data/Chatbot/quickstart#setup.txt']
index: [0, 1775, 0]
```

The function query_openai_and_faiss(store, query_text) converts the FAISS store into a document retriever with specific search configurations and executes a search in the FAISS index to find documents relevant to the query_text. It then constructs a context text from the retrieved documents, separating them with markers for clarity.

In addition to retrieving content, we extract source and index metadata from the retrieved documents, formatting this information alongside the response for reference.

The ChatPromptTemplate formats the final prompt by injecting the constructed context and the original query before submitting the prompt to the ChatOpenAI model. The template instructs ChatGPT to summarize the answer using only the information returned in the context. It is possible to augment its knowledge with LangChain using our private data to provide answers, rather than relying solely on the context data.

If we intend to use both OpenAI and FAISS data in the querying process, we need to integrate the embeddings generated by OpenAI with those stored in FAISS. This involves concatenating the embeddings from both sources and performing a similarity search on the combined set of embeddings, allowing for results from both OpenAI and FAISS data to be returned. However, alternative approaches to querying both OpenAI and FAISS data may provide more nuanced results or better performance. One such approach is to perform separate queries to OpenAI and FAISS and then combine the results using techniques like fusion or reranking.

It is perhaps worth clarifying the point on encoding and retrieval using a vector store. In a retrieval-based system like RAG (retrieval-augmented generation), the encoding process used during retrieval must be consistent with the encoding process used during indexing to ensure compatibility and effectiveness.

The two processes are conceptually distinct stages within the overall information retrieval pipeline. However, for the system to function effectively, the encoding process used during retrieval must align with the encoding process used during indexing.

In the RAG framework, documents are encoded into dense embeddings using a retrieval vectorizer during indexing. These embeddings are then stored in a retrieval index (e.g., a FAISS index) for efficient retrieval. During retrieval, queries are also encoded into dense embeddings using the same retrieval vectorizer. These query embeddings are then used to retrieve relevant documents from the index based on their similarity to the query embeddings. Therefore, it is crucial that the retrieval vectorizer used during retrieval is the same as (or compatible with) the one used during indexing. This ensures that both queries and documents are represented in the same semantic space, allowing for accurate matching and retrieval of relevant documents.

8.10 Other AI Agent Software

We have outlined a basic understanding of what AI agent software entails, defining it as programs capable of performing tasks or services for users autonomously or semi-autonomously. These software agents are underpinned by advanced artificial intelligence technologies, such as machine learning, natural language processing, and robotic process automation.

The landscape of AI technology is rapidly evolving, with developments progressing at an unprecedented pace. Innovations are continually being introduced, with both open source and proprietary models expanding the capabilities and accessibility of AI agents. A notable development in this space is Google's AI Agent Builder platform. This platform includes a "model garden" featuring a range of popular tools for AI agent development, such as Gemini, Claude, Llama, Stable Diffusion, and PaLM 2. These tools are designed to streamline the creation of more sophisticated and user-friendly AI agents.

The trend in AI development is shifting toward creating models that are increasingly intelligent and simpler for end users to interact with. This evolution means that intricate prompt engineering—once a critical aspect of programming AI—may become less essential. Instead, the emphasis is likely to shift toward enhancing the creativity and practical applications of AI technologies. We can expect to see significant advancements in various types of AI agent software, such as more adept virtual personal assistants, enhanced customer service bots, smarter business management tools, and more effective robotic process automation systems.

These next-generation AI agents are expected to exhibit human-like capabilities, delivering impressively contextual and nuanced responses across vast datasets. They will likely support natural speech interactions in multiple languages, further enhancing their usability and effectiveness. This progression points toward a future where AI agents not only replicate human tasks but do so with a level of sophistication that closely mimics human interaction and problem-solving abilities.

8.11 Concluding Remarks

I trust this book has fulfilled its intended purpose—to provide readers with a foundational understanding of programming and training deep neural networks. While we have touched upon numerous applications and models, the intricacies of some remain beyond the scope of this book due to their complexity. Nevertheless, readers should now possess the ability to comprehend the concepts and implementations of these advanced topics.

As with any technical discipline, true mastery comes from hands-on practice. Even the simplest coding exercises can impart invaluable insights that transcend the confines of a book. My aspiration is that this book has sparked curiosity within readers, inspiring them to embark on their own coding journeys. For it is through experimentation and exploration that one truly solidifies their understanding and hones their skills.

A legitimate question to ask is if much of these technologies are still relevant given the advent of generative AI. My view is that learning about older technologies, like support vector machines (SVMs) and generative adversarial networks (GANs), remains important and valuable, even in the era of advanced models like ChatGPT, for several reasons:

- Fundamental Understanding: Learning these technologies provides a solid foundation in machine learning principles. SVMs, for example, offer insight into classification, regression, and support vector concepts, while GANs demonstrate principles of unsupervised learning, distribution generation, and the innovative concept of adversarial training. This foundational knowledge is crucial for understanding how more complex models work.
- Applicability and Efficiency: Not all problems require the firepower of large models like ChatGPT. SVMs are highly effective for certain classification and regression problems, especially when dealing with small- to medium-sized datasets. They can be more computationally efficient and easier to deploy in resource-constrained environments. Similarly, GANs have unique capabilities in generating new data instances that can be invaluable for tasks like image generation, data augmentation, and more.
- Creativity and Innovation: Learning about a wide variety of technologies encourages creativity and innovation. By understanding the strengths and limitations of different approaches, researchers and practitioners can invent new models and algorithms that address the shortcomings of existing ones. Many advancements in AI come from combining ideas from different areas in novel ways.
- Customization and Control: Models like SVMs offer a level of transparency and control that is not always available in larger, more complex models. For tasks requiring explainability and the ability to fine-tune models based on specific criteria, these "older" technologies can be more suitable.
- Research and Development: Ongoing research in fields like SVM and GAN can lead to significant improvements and breakthroughs in machine learning. For

instance, GANs continue to be a hot area of research for generating synthetic data, with applications in training models where real data is scarce or privacy-sensitive.

In this dynamic landscape of technological advancement, it is essential to recognize the value of continuous learning and adaptability. The journey through the realms of programming, deep learning, and machine learning is not a linear path but a cyclical process of learning, applying, and relearning. As new models and techniques emerge, the foundational knowledge you've acquired from this book and beyond will serve as the bedrock upon which you can build, understand, and innovate.

The future of AI and machine learning is incredibly bright, and its potential applications are virtually limitless. From enhancing healthcare and advancing scientific research to improving environmental conservation efforts and revolutionizing industries, the possibilities are as vast as our collective imagination. As you stand at the threshold of this exciting frontier, remember that each challenge you encounter is an opportunity for growth, and every failure is a stepping stone to success.

Embrace the journey ahead with an open mind and a resilient spirit. Engage with the community, share your discoveries, and learn from the experiences of others. The field of AI is not just about algorithms and data; it is about the collective effort of individuals around the globe striving to make a positive impact through technology.

As we conclude this book, I hope it has not only imparted knowledge but also ignited a passion for exploring the uncharted territories of AI and machine learning. The path ahead is yours to shape. May you embark on this journey with confidence, curiosity, and a relentless drive to contribute to the ever-evolving tapestry of technological innovation.

Thank you for allowing me to be a part of your learning journey. Here is to the endless possibilities that await as you apply, challenge, and extend the knowledge you have gained.

Bibliography

1. Dumoulin, Vincent, and Francesco Visin. 2018. A guide to convolution arithmetic for deep learning.
2. Xue, Yanping, Rongguo Zhang, Yufeng Deng, Kuan Chen, and Tao Jiang. 2017. A preliminary examination of the diagnostic value of deep learning in hip osteoarthritis. *PloS one* 12: e0178992.
3. Sugawara, Yusuke, Sayaka Shiota, and Hitoshi Kiya. 2019. Checkerboard artifacts free convolutional neural networks. *APSIPA Transactions on Signal and Information Processing* 8: e9.
4. Liang, Feynman. 2016. BachBot: Automatic composition in the style of Bach chorales Developing, analyzing, and evaluating deep LSTM model for musical style. M.Phil Thesis. University Of Cambridge.
5. Chen, Yang,Yu-Kun Lai, and Yong-Jin Liu. 2018. CartoonGAN: Generative adversarial networks for photo cartoonization. In *Proceedings of the IEEE Conference on Computer Vision and Pattern Recognition*. Open Access Version.
6. Heaton, Jeff. https://github.com/jeffheaton/pyimgdata
7. https://stable-diffusion-art.com/how-stable-diffusion-work/
8. Vahdat, Arash, and Karsten Kreis. (2022). Improving Diffusion Models as an Alternative To GANs, Part 1. NVIDIA Techical Blog.
9. Radiuk, Pavlo. Applying 3D U-net architecture to the task of multi-organ segmentation in computed tomography. *Applied Computer Systems* 25: 43–50. Open Access.
10. https://gymnasium.farama.org/content/basic_usage/
11. https://platform.openai.com/docs/guides/images/usage
12. Boesch, Gaudenz. 2021. Very Deep Convolutional Networks (VGG) essential guide.
13. https://python.langchain.com/v0.2/docs/tutorials/rag/
14. https://python.langchain.com/v0.1/docs/modules/data_connection/vectorstores/
15. https://python.langchain.com/v0.1/docs/modules/data_connection/retrievers/

P. Hua, *Neural Networks with TensorFlow and Keras*,
https://doi.org/10.1007/979-8-8688-1020-6

Index

P. Hua, *Neural Networks with TensorFlow and Keras*,
https://doi.org/10.1007/979-8-8688-1020-6

T

U

V

Printed in the United States
by Baker & Taylor Publisher Services